The Magicians

Marcus Chown

M^THE^agicians

Great minds and the central
miracle of science

FABER & FABER

First published in 2020
by Faber & Faber Limited
Bloomsbury House
74–77 Great Russell Street
London WC1B 3DA

Typeset by Faber & Faber Ltd
Printed and bound by CPI Group (UK) Ltd, Croydon, CR0 4YY

A CIP record for this book
is available from the British Library

ISBN 978–0–571–34638–7

FSC
www.fsc.org
MIX
Paper from
responsible sources
FSC® C020471

2 4 6 8 10 9 7 5 3 1

To Manjit,
With love, Marcus

There are two kinds of geniuses: the 'ordinary' and the 'magicians'. An ordinary genius is a fellow whom you and I would be just as good as, if we were only many times better. There is no mystery as to how his mind works. Once we understand what they've done, we feel certain that we, too, could have done it. It is different with the magicians. Even after we understand what they have done it is completely dark.

MARK KAC
Genius: Feynman and Modern Physics
BY JAMES GLEICK

Contents

Introduction

The central magic of science

The universe is full of magical things patiently
waiting for our wits to grow sharper.
EDEN PHILLPOTTS[1]

Nothing is too wonderful to be true.
MICHAEL FARADAY

About 3.6 million years ago, three hominins walked across a volcanic landscape and left footprints in the recently fallen ash. The impressions of those footprints, which are visible today at Laetoli in Tanzania, are intensely evocative. As the biologist Richard Dawkins remarked, 'Who does not wonder what these individuals were to each other, whether they held hands or even talked, and what forgotten errand they shared in a Pliocene dawn?'[2]

We will, of course, never know the answers to these questions, but we can hazard a guess at some of the things the three hominins, probably *Australopithecus afarensis*, saw and wondered about on that distant day, long before the dawn of our own species. Much of the natural world is chaotic and unpredictable, but some things are regular and reliable: the rising and setting of the Sun; the march of the seasons; the changing phases of the Moon; the gradual drift of star patterns across the night sky. Such natural rhythms would almost certainly have left a deep impression on even our earliest ancestors.

There was no progress made in understanding these rhythms for tens of thousands of centuries after the footprints at Laetoli were made. Everything changed, however, with a critical

I

invention in the Middle East around 3000 BC: writing provided the means to record events in the sky and to recognise ever more subtle patterns in the movement of the heavenly bodies. In Babylon in modern-day Iraq, it became possible to predict astronomical spectacles such as eclipses of the Moon and the Sun. And those who made such predictions and controlled the dissemination of such information were able to strike awe in the minds of the population. Even if they were not tempted to pass themselves off as gods, they gained immense power over the masses.

That power, however, was nothing compared with the power of science. Science, which was born in the seventeenth century, found the ultimate reason for the world's patterns – the general 'laws' that underpin the rhythms of nature. And those laws are portable. So, although Isaac Newton famously deduced his law of gravity from the fall of an apple and the motion of the Moon around the Earth, he was also able to apply it in another, entirely different domain to explain why there are two tides in the oceans every twenty-five hours.[3]

Recognising a pattern in, for instance, eclipses permitted only the prediction of future eclipses. But science, by exploiting general-purpose laws, could predict the existence of phenomena that nobody had ever suspected. The first, and most striking, example of this was the prediction of an unknown planet by Urbain Le Verrier. When Neptune was found in 1846 – within a whisker of where in the night sky the French astronomer's calculations revealed it should be – it created an international sensation and made Le Verrier a superstar. 'Science has made gods of men,' the French biologist Jean Rostand would later write.[4]

The discovery of Neptune was a dramatic demonstration of the central magic of science: its ability to predict the existence of things previously undreamt of which, when people went out and looked for them, turned out to actually exist in the real universe. This ability is so magical that even the exponents of science can often scarcely believe it. Famously, Albert Einstein did not believe two predictions of his own theory of gravity: black holes and the Big Bang. And when it came to a third prediction – gravitational waves – he vacillated, predicting their existence in 1916 and unpredicting them a year later, before predicting them again in 1936. They would eventually be discovered on 14 September 2015.

The central magic of science appears miraculous because nobody knows why it works. The predictions made by physicists arise from mathematical formulae, or 'equations', which are found to describe aspects of the universe. But nobody knows why such equations so perfectly describe the physical world or, to paraphrase the twentieth-century Austrian physicist Eugene Wigner, why mathematics is unreasonably effective in the natural sciences. Put simply, the universe has a mathematical twin that can be written on a piece of paper or scrawled across a whiteboard. But why it has such a twin is a huge mystery.

The importance of the central magic of science is that it is at the crux of why physics works. Physicists naturally want to understand why the principal tool they use in their working lives is so effective, and understanding why it works will conceivably tell us something very profound about our universe and why it is constructed the way it is.

In this book I will tell the stories of some of the people who have demonstrated the central magic of science. One striking thing is the difference in their approaches. The Scot James Clerk Maxwell was arguably the greatest physicist between Newton and Einstein. His thought processes were essentially like those of a normal human being, though of course a souped-up version; in his mind, he concocted mechanical models of phenomena such as electricity and magnetism using everyday objects like cogs and wheels. Only when he was satisfied that he had captured the essence of reality did he express his model in mathematical terms. In the case of electricity and magnetism, this yielded his famous 'equations of electromagnetism', which revealed that light is an 'electromagnetic wave' and predicted the existence of radio waves, making possible the ultra-connected world of the twenty-first century. The approach of the English physicist Paul Dirac, however, was very different: the hyper-literal 'Mr Spock of physics' simply plucked the formula which describes an electron travelling at close to the speed of light out of thin air. The 'Dirac equation', which predicted a hitherto unsuspected universe of 'antimatter' and is one of only two equations inscribed on the stone floor of Westminster Abbey, was the result of Dirac playing with equations on a piece of paper and insisting on mathematical consistency.

The stories I tell here of Maxwell, Dirac and many others who demonstrated the central magic of science are as factual as I can make them. If the scientists are alive and it was possible to interview them, I did so; for those that are dead, I used the facts at my disposal and dramatised the events around them. For instance, my description of the day that Maxwell came to the stunning realisation that light is a wave of electricity and magnetism is pieced together from the available facts. On his return from a summer holiday at his Glenlair estate in Scotland, he did indeed go to the library at London's King's College to look in a reference book for the measured values of the permittivity and permeability of air, which had been obtained by Wilhelm Weber and Rudolf Kohlrausch. He did walk or catch the horse-drawn bus from his home in Kensington to the Strand and back each day, a route that took him along Piccadilly and past the Albemarle Street turn-off to the Royal Institution, where he would sometimes stop. And he and his wife did ride regularly in Hyde Park and Kensington Gardens, Katherine's pony Charlie having been brought down to London by train from Glenlair.

My hope is that, by dramatising such stories of scientific prediction and discovery, I will not only bring the events alive but also provide some idea of what the moment of discovery must be like and how exhilarating it must be to realise a profound truth about the world that no one has known before. For those interested in the history of science, I have provided copious references.

This is the story of the magicians who, with pen and paper, not only predicted the existence of unknown worlds, black holes and subatomic particles but antimatter, invisible waves that course through the air, ripples in the fabric of space–time and many more things besides. This is the story of the central magic of science and how it made gods of men.

1

Map of the invisible world

The hypotheses which we accept ought to explain phenomena which we have observed. But they ought to do more than this: our hypotheses ought to foretell phenomena which have not yet been observed.

WILLIAM WHEWELL[1]

I grew up believing my sister was from the planet Neptune and had been sent down to Earth to kill me.

ZOOEY DESCHANEL

Berlin, 23 September 1846

They had been searching for almost an hour and had already slipped into an automatic rhythm. Johann Galle squinted through the giant brass refractor at the clear night sky, adjusted the controls of the telescope until a star appeared in the cross hairs and barked out its co-ordinates. His young assistant Heinrich d'Arrest was seated at a wooden table across the stone floor of the observatory dome. He ran his finger over a star map by the light of an oil lamp and shouted back, 'Known star.' Galle twiddled the brass knobs again, lining up another star. Then another. In the chilly night air, he was already getting a crick in his neck and was beginning to wonder whether they were wasting their time.

The director of the Berlin Observatory, Johann Franz Encke, had certainly thought so that afternoon when Galle had appeared at his office door with his unusual request. But because Encke planned to be at home that night celebrating his fifty-fifth birthday rather than at the 22-centimetre refractor, he had given Galle permission to use the instrument.

The exchange between Galle and Encke had been overheard by d'Arrest, an astronomy student who was lodging in one of the observatory's outbuildings so he could gain more practical experience; he immediately begged Galle to let him help. And so here the pair of them were, on the crystal-clear night of 23 September 1846, scanning the skies with the great, clock-driven Fraunhofer telescope, one of the most advanced instruments of its kind in the world.

They had started their search when the gaslights of Berlin sputtered off, plunging the city into blackness, and it was now approaching midnight. Galle manoeuvred the cross hairs to the next star and called out its co-ordinates. His mind wandered to thoughts of the warm bed he would soon be sharing with his wife and he began to think how ridiculous he would seem in the morning when he told Encke of their failure. He waited for a response from d'Arrest. And waited. What in the world, he wondered, was his assistant doing?

The crash of a chair hitting the floor shocked Galle back to reality. Leaping back from the eyepiece, he saw his assistant silhouetted against the oil lamp, rushing towards him, flapping his star map like a demented bird. It was too dark to make out the expression on d'Arrest's face, but Galle would remember his words for the rest of his life: 'The star is not on the map! It is not on the map!'

Paris, 18 September 1846

The man who had suggested looking for a star that was not on any star chart, in a letter that had arrived at the Berlin Observatory on 23 September, was Urbain Le Verrier. An astronomer at the École Polytechnique in Paris, Le Verrier was interested not in observing celestial bodies from draughty telescope domes but in sitting at his desk and using Newton's law of gravity to calculate the orbits of such bodies and compare them with existing observations. In the course of this work, he had become obsessed by a planet which seemed to break all the rules: Uranus.

Uranus had been discovered by a musician from Hanover in Germany. In 1757, William Herschel, aged just nineteen, had moved with his sister Caroline to Bath in the west of England, a pretty spa town which had been developed by the Romans because of its hot springs. He found work as a church organist, but his real passion was astronomy, and in the garden of his house he built one of the best telescopes of his day. It was on 13 March 1781, while scanning the night sky with this instrument, that a fuzzy star popped into his eyepiece. At first Herschel thought it was a comet, but unlike a comet it did not have a gossamer tail. Not only that but, as it crept across the constellation of Gemini over the subsequent nights, it did not follow the highly elongated orbit of a comet but the near-circular orbit of a planet.

Herschel had discovered the first new planet in the age of the telescope, the first world entirely unknown to ancient astronomers. Throughout all of recorded history, the number of planets had stood at six. Now, incredibly, there were seven. Herschel's discovery created an international sensation and elevated him to the status of a scientific superstar.

Herschel's greatest desire, as an immigrant, was to be accepted by his adopted country, and he therefore christened the new planet 'George', after King George III (actually, he named it 'George's star'). Not surprisingly, French astronomers objected to having a planet named after an English king and instead referred to it as 'Herschel'. In an attempt to make the peace, the astronomer Johann Bode suggested it be named after Uranus, father of the Roman god Saturn, and the name stuck. (If it had not stuck, the planets, in increasing distance from the Sun, would have been Mercury, Venus, Earth, Mars, Jupiter, Saturn . . . and George.)

Actually, Uranus had been seen almost a century earlier, in 1690, by the English astronomer John Flamsteed, but he had mistakenly believed it to be a star and catalogued it as 34 Tauri, the thirty-fourth star in the constellation of Taurus. Historical records of the planet's position were able to supplement new observations of the planet; consequently, by the early nineteenth century, its orbit was known precisely enough that it could be compared with that predicted by Newton's law of gravity. But the comparison threw up an anomaly.

Whenever an orbit was predicted for Uranus, the planet would stray from it over the following months. No one seriously believed there was anything wrong with Newton's law of gravity. Its successes had been so overwhelming and so comprehensive that it had a status akin to the word of God. Instead, the suspicion arose that Uranus was constantly veering from its calculated orbit because it was being tugged by the gravity of another world even further from the Sun. It was a tantalising possibility, and Le Verrier could not resist the challenge of pursuing it. Seated at his desk at the École Polytechnique in Paris, he set out to deduce, from the observed effect of the hypothetical planet on Uranus, exactly where in the night sky it must be.

The Sun accounts for a whopping 99.8 per cent of the mass of the solar system, so, to a very good approximation, a planet can be assumed to be moving under the influence of the Sun alone. However, Newton's law of gravity is a 'universal' law, which means there is a force of attraction between every chunk of matter and every other chunk of matter; consequently, a planet is influenced not only by the gravitational tug of the Sun, but of all the other planets as well. To be sure he was seeing the effect of an unknown planet in the outer solar system on Uranus, Le Verrier needed to first subtract the effect of the known planets, and especially of the two most massive ones: Jupiter and Saturn.

The calculations were complex and time-consuming. Each one had to be checked and re-checked, since a single small error might be magnified and bring the whole mathematical edifice crashing down. But that was not the only problem Le Verrier faced: the gravitational pull of a lightweight planet close to Uranus would be indistinguishable from that of a massive planet far away. Therefore, in order to make any progress in pinning down the orbit of the hypothetical planet, Le Verrier had to guess its mass and distance from the Sun.* It was a gargantuan task that took up all of his working days and some of his nights as well. But, eventually, Le Verrier succeeded. He deduced not only an orbit for the hypothetical planet but, most importantly, where in the night sky a telescope should be

* In guessing the distance of the hypothetical planet from the Sun, Le Verrier was aided by the Titius–Bode law, although no scientific reason is known why the planets should follow such a rule. See http://demonstrations.wolfram.com/TitiusBodeLaw/.

pointed to look for it: between the constellations of Capricorn, the goat, and Aquarius, the water carrier.

Le Verrier was a confident man but, as his quill hovered above the dense formulae that covered the pages spread across his desk, he felt a buzz of nervous excitement. To know something that no one else in the world knew or understood was a most exhilarating feeling of power, but could he be wrong? Was he a god, or merely a fool? And how was it possible that the equations before him described reality? Before he could be overwhelmed by doubts, he pulled himself together. There was only one thing for him to do: inform the observational astronomers.

Le Verrier took the location of the new planet to the director of the Paris Observatory, but François Arago made it clear that he did not think the search for a new planet was a priority. He had good reason. For a start, national observatories such as his in Paris existed principally to make charts of the locations of planets and stars for the purposes of navigation. This involved many people carrying out lengthy and painstaking observations, and Arago understandably did not want to use up their valuable time on a wild goose chase for a planet whose existence seemed to him to be the remotest of long shots. It probably did not help that Le Verrier was a man with a reputation for being arrogant and difficult to deal with.

Capricorn and Aquarius would not be visible from the Northern Hemisphere much later than November, so it was imperative that any search for the new planet begin soon. For a while Le Verrier was patient, but when a start date from Arago was not forthcoming, his frustration grew. As it happened, he had already started trying other avenues and had sent a paper containing his predictions to the editor of the

German journal *Astronomische Nachrichten*. In his accompanying letter to Heinrich Schumacher, he vented his frustration
at not being able to get French astronomers to look for his
planet. Schumacher was sympathetic and wrote back with a
suggestion: why did Le Verrier not contact other astronomers
with powerful telescopes? The two that came immediately to
his mind were Friedrich Struve in Germany and Lord Rosse at
Birr in Ireland, whose 'Leviathan', with its 72-inch mirror, was
the biggest telescope in the world. Le Verrier would probably
have contacted both of them had Schumacher's suggestion not
reminded him of a letter he had received the previous year from
a young astronomer at the Berlin Observatory.

The appeal of Johann Galle was that he was a low-ranking
assistant astronomer. Le Verrier expected that Johann Encke,
director of the Berlin Observatory, would be as reluctant to
search for a new planet as his Parisian counterpart, but Galle
might be hungry to make his name. Le Verrier might have more
success, he reasoned, by bypassing Encke and contacting the
younger astronomer directly. Would Galle take him seriously,
or would Le Verrier be disappointed yet again? There was only
one way to find out.

The only problem was that the French astronomer had
ignored a letter from Galle a year earlier, together with a thesis
which had been included with it. This was embarrassing now
that he needed a favour from him. However, a bit of flattery
might get around that, so before he made his request that
Galle embark on a search for his planet, Le Verrier penned
some pointed and belated praise, congratulating Galle on the
'perfect clarity' and 'complete rigour' of his thesis. Then, on 18
September 1846, he sent his letter, which contained a rough
estimate of the location of the new planet, to Berlin.

Berlin, 24 September 1846

As the clock ticked towards dawn, three men were gathered at the Fraunhofer telescope in the dome of the Berlin Observatory. D'Arrest, who had run all the way to Encke's home, had returned with the observatory director, who was a little unsteady on his feet after his birthday celebrations. The trio, struggling to stay calm, took turns at the eyepiece until they were absolutely sure. The object seen by Galle and d'Arrest was definitely not on the star map. And the reason was patently clear: it was not a star. Stars, because of their distance from Earth, appear as pinpricks of light no matter how powerful the magnification of a telescope. But this object was no dimensionless pinprick. It was a tiny shimmering disc. They had found it! They had found Le Verrier's planet!

Galle could scarcely believe the events of the last half-day. He had knifed open what looked like a perfectly ordinary letter from France, not for an instant suspecting that it would change his life forever. He recognised Le Verrier's name immediately and could easily have exacted his revenge on the Frenchman for ignoring him by losing his letter among the papers on his desk. But the favour Le Verrier had requested piqued his interest.

The letter contained a prediction of the existence and location of a new planet. Galle knew that such a prediction was ridiculous, yet something caused him not to dismiss it out of hand. 'I would like to find a persistent observer', wrote Le Verrier, 'who would be willing to devote some time to an examination of a part of the sky in which there may be a planet to discover.' Galle decided to be that persistent observer.

If the truth be admitted, Galle had never expected to find anything. It did not seem possible. How could a man sitting at a desk in Paris 'see' the universe with the aid of mathematics? It was about as likely as a blindfolded astronomer discovering a comet with the Fraunhofer telescope. But – miracle of miracles – there it was: Le Verrier's planet, looming out of the inky depths of space, exactly where the man had predicted it would be.

The new world had been trailing around the Sun in the frigid darkness beyond the orbit of Uranus since the birth of the solar system. And until an hour ago no human being had known of its existence. For the moment, they were the only three people on Earth who had seen it, and it had no name. Soon, however, everyone in the world would know it as Neptune.

Paris, 29 September 1846

In Paris, a few days later, Le Verrier tore open a letter from Berlin, dated 24 September 1846. 'Sir,' he read. 'The planet whose position you have pointed out actually exists.'

Galle had found his planet! Le Verrier was dizzy with euphoria, but also relief. He had believed in the new planet – of course he had – but he had also not believed in it. He was human, after all. He had staked his reputation on a piece of arcane mathematics, which the Creator may or may not have decided to respect. He had sounded confident when he made his prediction, but he alone knew how much of that was bravado.

On 1 October, Le Verrier replied to Galle. He thanked the German astronomer profusely for being the only one to take seriously his request, writing, 'We are, thanks to you, definitely in the possession of a new world.'

The discovery of Uranus had been a sensation: twice as far from the Sun as Saturn, it had overnight doubled the size of the solar system. But the discovery of Neptune was a sensation of an entirely different order. Whereas Herschel had stumbled on Uranus by accident, the existence of Neptune, its location and even its appearance had been predicted by Le Verrier, with nothing more than a pen and paper.

'Without leaving his study, without even looking at the sky,' wrote French astronomer Camille Flammarion, 'Le Verrier had found the unknown planet solely by mathematical calculation, and, as it were, touched it with the tip of his pen!'[2]

Discovering something in the real world from a desk, as Flammarion recognised, was truly something new under the sun. 'The entire annals of Observation probably do not elsewhere exhibit so extraordinary a verification of any theoretical conjecture adventured on by the human spirit!' wrote the Scottish astronomer John Pringle Nichol.[3]

But the discovery of Neptune was a triumph not only for Le Verrier; it was also a triumph for Isaac Newton and the universal theory of gravity he had devised almost two centuries earlier. Newton's law explained not only what we could see but predicted what we could not.

Le Verrier had demonstrated in spectacular fashion the central magic of science – its astonishing ability to predict things never before suspected which turn out to exist in the real world. It stretched credulity that mathematical equations scrawled across a page could so perfectly capture reality, but miraculously, they did. Using abstract formulae, Le Verrier had uncloaked a real body in the real world, and nobody in the history of the world had done anything like it. Le Verrier was the first of the magicians.

o o o

The discovery of Neptune triggered a heated priority dispute between France and England because an English mathematician had also used the anomalous motion of Uranus to predict the location of the new planet. John Couch Adams was an autistic mathematical genius from Cornwall in England. In 1841, while a student at the University of Cambridge, he set out to deduce where in the night sky the new planet must be in order to have the observed effect on Uranus. His calculations took four years, but in 1845, he took his result to Sir George Biddell Airy, Astronomer Royal and director of the Royal Observatory at Greenwich. Unfortunately, like Le Verrier in France, he was fobbed off. When Airy did eventually take notice of Adams, rather than publicising his prediction and authorising a search with one of the Greenwich telescopes, he chose to pass the information to George Challis, who had succeeded him as director of the Cambridge Observatory.

Challis could immediately see that Adams' prediction was not a precise location but an extended patch of the sky where the hypothetical planet might be found. A comprehensive search would take almost one hundred sweeps with the Cambridge transit telescope, each of which would last several hours. Estimating that the whole process would take around 300 hours of observing time, Challis held off for a while. When he eventually started the search, he recorded Neptune – twice, in fact – while failing to recognise it. By then it was too late and Galle in Berlin had already found the new planet.

The episode was a great embarrassment for Airy and Challis, since they had received a prediction of the new planet's location from Adams before Galle received a similar one from Le Verrier. Things were made worse by the fact that they had kept

Adams' prediction a secret, perhaps to make sure that, if it was discovered, Cambridge would have the glory. However, the fact that none of Adams' calculations had ever been published made the French suspicious that there had ever been an English prediction.

The international dispute over Neptune was prolonged and bitter, but to the credit of Adams and Le Verrier, neither of them took any part in it. Perhaps because they appreciated each other's mathematical wizardry and had faced similar obstacles in getting mere mortals to take them seriously, as soon as they met in England they became firm, lifelong friends. Nowadays, as often as not, the discovery of Neptune is attributed jointly to Adams and Le Verrier.

o o o

After his triumphant prediction of the existence of Neptune, Le Verrier's star rose in the scientific firmament and in 1854 he became director of the Paris Observatory. But nothing he achieved came close to matching the exhilaration he had felt at magically unveiling an unknown world at the edge of the solar system. He had been courted by kings and revered as a god by scientists. Fame and adulation had intoxicated him, and craving that feeling again, he turned his attention from the outer to the inner solar system.

Le Verrier's goal was to understand thoroughly the orbits of the inner planets: Mercury, Venus, Earth and Mars. If he could do that, then perhaps, just perhaps, a Uranus-type anomaly would show up that would lead to a headline-grabbing discovery. Remarkably, such an anomaly did indeed exist, and it concerned the innermost planet. Even when the gravitational tug of the other planets on Mercury was taken into consideration, it still did not move as expected.

Le Verrier became convinced that there was a planet orbiting even closer to the Sun than Mercury and, by February 1860, it had a name. Planets are named after ancient gods, and the lord of the forge on Mount Olympus, home of the Greek gods, was Vulcan. It seemed an appropriate name, since the new world could never escape the fires of the Sun.

For almost half a century, astronomers searched for Vulcan, but gradually it fell out of favour and all sightings of it turned out to be mirages. The anomalous motion of Mercury remained, and nobody suspected what it was really telling us: that, incredibly, impossibly, Newton was wrong about gravity. Nobody, that was, until Albert Einstein, who in 1915 devised a better theory of gravity – the general theory of relativity – to supplant Newton's.

But although Vulcan had been a dead end, Neptune very definitely had not been. Le Verrier had shown how it was possible to use Newton's law of gravity to predict what we could not see – to make a map of the invisible world.

o   o   o

In the first decades of the twentieth century, there was a suggestion that the orbit of Neptune, just like that of Uranus, was being perturbed. It turned out to be untrue. Nevertheless, it triggered a search for a 'Planet X', even more distant from the Sun. This culminated on 18 February 1930 in the discovery of Pluto, the only planet to be named by a child – eleven-year-old Venetia Burney from Oxford.[4]

Pluto, which is smaller even than Earth's moon, turned out to be far too tiny to affect Neptune. In fact, at the end of the twentieth century, it was revealed to be one of tens of thousands of similar bodies circling the Sun beyond the orbit of Neptune. It was the discovery of this 'Kuiper Belt' of icy builder's rubble left over from the formation of the solar system 4.55 billion years ago that led the International Astronomical Union to demote Pluto in August 2006 from a planet to a 'dwarf planet'.

But Newton's law of gravity may not yet have exhausted its ability to reveal the invisible in our solar system. At the beginning of 2016, two astronomers at the California Institute of Technology in Pasadena pointed out that at least half a dozen Kuiper Belt Objects are moving oddly. Mike Brown and Konstantin Batygin claim that their motion is due to them being tugged by an unknown planet orbiting the Sun at the periphery of the solar system.[5] But rather than a celestial tiddler like Pluto, this planet would have about ten times the mass of the Earth.

Brown and Batygin claim 'Planet 9' orbits on average about twenty times as far from the Sun as Neptune. Since planets shine only due to reflected sunlight, it would be extremely faint and hard to find, but many astronomers are keen to be the new Johann Galle and searches for Planet 9 are already underway.

The real success of the technique pioneered by Adams and Le Verrier, however, has been in detecting the anomalous motion of stars caused by the gravitational tug of their invisible planets. In 1995, 51 Pegasi b was the first planet to be discovered around a normal star other than the Sun; now more than four thousand 'exoplanets' are known and the total is rising at an ever-quickening rate.

But arguably the most important invisible thing revealed by Newton's law of gravity is 'dark matter'. Although its existence was first suspected in the 1930s by Swiss–American Fritz Zwicky and Dutchman Jan Oort, it took the work of two astronomers at the Department of Terrestrial Magnetism of the Carnegie Institution of Washington to confirm its existence. In the late 1970s and 1980s, Vera Rubin and Kent Ford found that stars in the outer regions of spiral galaxies are orbiting the centres far too fast. Like children on a speeded-up merry-go-round, they should be flung into intergalactic space.

Astronomers have explained this anomaly by suggesting that there is much more matter in spiral galaxies than we see in the form of stars, and that it is the extra gravity provided by this invisible dark matter that holds on to the outermost stars. Across the universe, dark matter outweighs the visible stars and galaxies by a factor of about six. Nobody knows what it is made of, although the best bets are undiscovered subatomic particles or Jupiter-mass black holes left over from the Big Bang. If you can figure out the identity of the dark matter, there is a Nobel Prize waiting for you in Stockholm.

2

Voices in the sky

This velocity is so nearly that of light, that it seems
we have strong reasons to conclude that light itself
(including radiant heat, and other radiations if any)
is an electromagnetic disturbance in the form of
waves propagated through the electromagnetic field
according to electromagnetic laws.

JAMES CLERK MAXWELL

From a long view of the history of mankind – seen
from, say, ten thousand years from now – there can
be little doubt that the most significant event of
the nineteenth century will be judged as Maxwell's
discovery of the laws of electrodynamics.

RICHARD FEYNMAN[1]

Karlsruhe, Germany, 13 November 1887

Today was the day. He was sure of it. Heinrich Hertz bolted
down his breakfast, kissed his wife Elisabeth and his baby
daughter Johanna goodbye, and hurried through the streets of
Karlsruhe to the university campus. On reaching his laboratory, he pulled down the blinds and switched on the 'oscillator' circuit that he and his assistant, Julius Amman, had been
building these past few days. The current surged through the
20,000-volt induction coil and he heard a faint crackle but
could see nothing. Only when his eyes adjusted to the gloom
was he sure that a spark was stuttering in the 7.5-millimetre

air gap he had left in the circuit. Satisfied that his 'transmitter' was working as intended, he turned to his 'receiver'.

A metre and a half away along the bench, Hertz had propped up a vertical loop of copper wire, which also contained a tiny air gap. He adjusted the gap with a screw to make it as small as possible and squinted at it in the laboratory gloom. Nothing.

He returned to his transmitter. Because the frequency of his oscillator circuit was so high, the spark was leaping back and forth across the air gap too rapidly for him to detect any motion with the naked eye. On each side of the gap there was a 1.5-metre length of conducting wire, terminated by a thirty-centimetre-diameter zinc ball. By moving the zinc balls along the wires, Hertz could change the 'capacitance' of the circuit and, with it, the frequency of the leaping spark. He did this several times while peering closely at his receiver, which he had 'tuned' so that, if it felt a vibration at a particular frequency, it would oscillate in sympathy. Still nothing.

He moved the zinc balls along the wires by only a few millimetres at a time and worked like this all morning, steady, patient and unhurried. It was a joy to at last have his own laboratory, a luxury he had only dreamt of while at the University of Berlin, where until 1885 he had been the assistant of Hermann von Helmholtz, Germany's most famous scientist. He was also perversely grateful for the economic recession which had recently engulfed the country; although it had left the department he headed bereft of students, the side effect was that he could devote himself to his research.

Hertz made yet another adjustment and stood stroking his neat beard while the thought went through his mind: was this really going to work? But a subtle change in the sound in the laboratory stopped him in mid-stroke. Frowning, he crouched next to his receiver.

There was a spark in the air gap! The gap was only a few hundredths of a millimetre wide and the spark was easier to hear than it was to see, but there was no doubt about it. It was definitely there.

He switched off the oscillator and the spark at the receiver died. He switched it on again and it reappeared. Something invisible was travelling through the air from his transmitter to his receiver! Although he could not prove it yet, he was sure he knew what it was. It had been predicted fifteen years earlier by a brilliant Scottish physicist who had died tragically young.

London, October 1862

When he left King's College, James Clerk Maxwell felt like dancing on air. The autumn rain had stopped and the sun had come out, and he stopped opposite St Mary le Strand Church, utterly transfixed by the light sparkling on the surface of a puddle in the road. An hour ago, it had been only a suspicion in his mind. But now, having consulted a reference book in the library and plugged some numbers into his theory, it was a fact. He knew something that nobody in the history of the world had known before: he knew what light was.

The shout of a man on a hay wagon snapped him out of his reverie just in time to avoid his foot being crushed under a heavy cartwheel. He set off down the Strand, dodging the costermongers, flower sellers and vagrants. Although his usual habit was to walk the four miles from his home in Kensington to King's each morning and catch the horse-drawn bus home, today, because of his desire to get to the library as quickly as possible, he had caught the bus in and was walking back.

He passed through Trafalgar Square, walked along Pall Mall East, turned up Haymarket and eventually came to the broad thoroughfare of Piccadilly. He had intended to go straight home – he had promised his wife, Katherine, they would go riding in Hyde Park – but as he came to Albemarle Street, he felt a compulsion to turn down it. Leaving the hubbub of the busy street behind him, he headed towards the building with a neo-classical façade and giant Corinthian pilasters at the end of the road.

The Royal Institution was where Michael Faraday had carried out his groundbreaking experiments on electricity and magnetism, and where the great man had instituted his Christmas Lectures for children and adults in 1825. Maxwell himself had also lectured there many times since his move to London from Aberdeen in 1860. During one triumphant lecture in May the previous year, he had even projected onto a big screen an image of a tartan ribbon – the world's first colour 'photograph'.[2]

Faraday was forty years Maxwell's senior and an old man of seventy-one. Four years earlier, with his health failing, he had retired to Hampton Court, on the river to the west of London. He still made occasional visits to the Institution, and Maxwell had hoped that he might be fortunate enough to catch his friend there and share his discovery, but he was not in luck. With the permission of the Institution's doorman, Maxwell went downstairs to the basement. In Faraday's abandoned magnetic laboratory, he surveyed the coils and batteries and bottles of chemicals gathering dust. Without the experiments Faraday had carried out here, Maxwell knew that his remarkable discovery would have been impossible.

o   o   o

Faraday's beginnings could not have been more different from Maxwell's. Maxwell was heir to the 1,500-acre country estate of Glenlair in the Vale of Urr, near Dumfries in southern Scotland; it was from there that he had returned to London by train a day ago. Faraday, by contrast, was the son of a poor blacksmith.* At fourteen, he had been apprenticed to a bookbinder in Marylebone, just off Oxford Street, the route along which until only a few decades earlier condemned prisoners had been taken by cart from Newgate Prison to the gallows at Tyburn.

George Ribeau, a Huguenot refugee, had encouraged his apprentice to read the books that he was binding, many of which were scientific. In a bid to educate himself further, Faraday had attended weekly lectures at the City Philosophical Society, which were given by the society's founder, the silversmith John Tatum, at his home in nearby Dorset Street. Inspired by the idea that he should believe only things he could demonstrate himself, Faraday started his own scientific experiments with what equipment he could afford on his meagre wages. He also made beautifully illustrated notes of Tatum's lectures. These proved of crucial importance in getting him his life-changing break when Ribeau showed them to a client in his shop at 48 Blandford Street.

George Dance, an architect and artist, asked whether he could show Faraday's notes to his father, a member of the Royal Institution. The next day, he returned to the shop with a ticket to a series of lectures by Humphry Davy. Like the golden ticket

---

* Surprisingly, Jacob's Well Mews, close to Marylebone High Street, has no plaque to mark the site of the forge, despite its importance in Faraday's life.

in Roald Dahl's *Charlie and the Chocolate Factory*, the gift would turn out to be a passport to a better life for Faraday – though not immediately.

Davy was the most famous British scientist of his day, a man who had invented the miners' safety lamp, discovered numerous new elements and who lectured with the showbiz razzmatazz of a music hall star.[3] Half his audience were women, who reportedly swooned at his dashing presence. Faraday could barely contain his excitement when the evening of the first lecture came and he found himself among a chattering high-society crowd queuing beneath the flickering braziers of the Royal Institution.

In 1812, with his apprenticeship with Ribeau nearing its end, the twenty-one-year-old Faraday took up a professional post, resigned to a future of bookbinding drudgery. But he had a piece of good fortune when Davy was temporarily blinded by an explosion in his laboratory and Dance Senior suggested that Faraday might help him; for a few euphoric days, he became his hero's assistant.

Afterwards, Faraday was afraid that he might never experience the scientific life again. However, he had an idea, and, using the skills he had acquired during his apprenticeship, he bound the notes he had taken during the Royal Institution lectures and sent them to Davy. It was a long shot, but on Christmas Eve he received a reply, promising him an interview in the New Year. The interview happened, but Faraday was plunged back into gloom when Davy said he had no vacancy.[4] Then, one day, there was a miracle. A carriage drew up outside the Faradays' house and a footman got down with a letter from Davy. He had fired his bottlewasher for fighting. The job, if he wanted it, was Faraday's.

Davy was by this time the greatest scientist in Europe. His native country had knighted him, and France so revered him that it had awarded him the Napoleon Prize, even though it was at war with Britain. But Davy's greatest success would turn out to be Michael Faraday.

Both Davy and Faraday, who eventually became his assistant, were fascinated by electricity. Davy had pioneered the field of 'electrochemistry', the technique by which he had isolated nine chemical elements, including potassium, sodium, calcium, barium, strontium and magnesium.

At the beginning of the nineteenth century, electricity was at the forefront of science and the popular imagination. It seemed so mysterious and unearthly that some even considered it satanic. Luigi Galvani's discovery in around 1781 that electricity could twitch the leg of a dead frog had inspired the precocious eighteen-year-old Mary Shelley to write *Frankenstein* in 1818.[5] But the most significant development of the time was Alessandro Volta's invention of the battery in 1799; by generating a continuous current, it made possible the scientific study of electricity.[6]

However, it was the news of a sensational discovery in Denmark that caused Davy and Faraday to drop everything. On 21 April 1820, Hans Christian Ørsted was lecturing at the University of Copenhagen when he noticed that a compass needle was deflected from magnetic north whenever he switched the electric current in a nearby wire on or off. The needle was deflected exactly as it would have been had it been close to a magnet; the unavoidable conclusion was that a current-carrying wire was a magnet. Might this discovery even explain why some materials such as iron were magnetic? Might electric currents be circulating deep inside those materials? No one had guessed it before, but there was a connection between electricity and magnetism.

On 4 September 1821, Faraday used the effect discovered by Ørsted in an ingenious way.[7] In his basement magnetic laboratory, he arranged a current-carrying wire so that it was continually deflected by a fixed magnet and so circled endlessly. It was not a very practical arrangement – it involved a bath of conducting mercury, which was highly toxic – but it proved the principle of the electric motor. Actually, Faraday had created the world's first electric motor the day before, using a fixed wire and an endlessly circling magnet rather than a fixed magnet and an endlessly circling wire.

Maxwell would have loved to have witnessed that magnet circling under the influence of a mysterious but invisible force while horse-drawn carriages rumbled past on Albemarle Street. It must have seemed as if some impossible wonder from the distant future had fallen into nineteenth-century London through a crack in time. Faraday had been with his fourteen-year-old nephew George, and the pair of them had been so euphoric at the sight of the endlessly circling magnet that they had danced around the laboratory table, before heading off to a circus to celebrate.

The obvious question was: If electricity could create magnetism, could magnetism create electricity? It took Faraday until the summer of 1831 to find the answer, by which time Davy had died and he had succeeded him as Director of the Royal Institution.

Shortly after Ørsted's discovery that a current-carrying wire behaves like a magnet, the French scientist André-Marie Ampère, 'the Newton of Electricity', had found it was possible to boost the effect by creating a cylindrical spiral out of wire.[8]

The more turns of wire in such a 'solenoid', the more powerful its magnetic effect. The proviso was that neighbouring sections of wire should not touch each other so that electricity leaked between them, which required interposing 'insulating' materials that did not conduct electricity.

Faraday turned to a solenoid in his attempt to use magnetism to create electricity. Iron was known to greatly enhance the magnetism of a solenoid, so he used it in the form of a ring that was fifteen centimetres in diameter. Around either side of the ring he wound a tight spiral of wire. Between each turn of the coil and its neighbour he interposed lengths of string, and he used sheets of cloth to insulate each layer from the next and from the iron ring. Although the two solenoids were physically unconnected, Faraday expected that when an electric current went through the first coil, turning it into a magnet, its magnetic tendrils would reach through the air to the second solenoid.

Faraday flipped a switch, sending an electric current through the first solenoid; to his delight, a current appeared fleetingly in the second coil. He then switched off the current in the first solenoid and a current appeared in the second coil – this time, bafflingly, flowing in the opposite direction. It was an epoch-making discovery: he had succeeded in making electricity from magnetism.

Faraday later found an easier way to achieve the same end, by simply inserting a bar magnet into the coil of a solenoid. When he pushed it in, a current flowed one way, and when he pulled it out, it flowed the other way. Faraday did not know it, but his discovery of 'electromagnetic induction' would change the world, leading to the development of 'dynamos' capable of the large-scale generation of electrical power.

That electricity and magnetism were connected was now beyond any doubt, but the fundamental questions remained. What was electricity? And what was magnetism? Although these mysteries continued to tantalise Faraday, his ground-breaking experiments had given him a feel for how electricity and magnetism worked, which led him to entertain a radical – in fact, heretical – idea.

When Faraday held a piece of iron close to a magnet, he could feel the magnetic force of attraction reaching out to grab it and concluded that there must be something invisible but real in the air in the space around it. And when he rubbed a piece of amber with fur, 'charging' it with 'static' electricity, it grabbed tiny scraps of paper, leading him to believe that there was something invisible but real in the air around the electric charge.

In Faraday's mind, a magnet set up a magnetic force 'field' around it, and it was this that acted on a piece of metal. Similarly, an electrically charged body set up an electric force field in the space around it, and it was this that acted on the scraps of paper. Faraday imagined he could almost see the fields, like a wind or swirling fog, permeating empty space.

In perceiving the world in this way he was completely alone. At the time, everybody thought that electric currents were the important thing, but Faraday was sure that the fields were key. To his mind, a conductor was merely a guide for an electric field, which existed in the space around the wire and was the principal carrier of energy. An electric current was merely a sec-ondary effect, a flow of electric 'charge' urged on by the electric field where it happened to intersect the conductor.

The field idea revealed the pleasing symmetry in the dis-coveries of Ørsted and Faraday. Ørsted's discovery that a current-carrying wire was a magnet showed that a changing electric field creates a magnetic field, and Faraday's discovery of electromagnetic induction showed that a changing magnetic field generates an electric field.

The reason Faraday's idea of the field was shocking and heretical was because of the success of Isaac Newton. The greatest scientist in history had been spectacularly successful in explaining another fundamental force – the force of gravity – as acting instantaneously across space. According to Newton's 'universal theory of gravity', the gravitational effect of the Sun acts directly on the Earth, and there is no medium through which the force is transmitted. This idea of 'instantaneous action at a distance' is, of course, ludicrous. Newton himself said as much; it was just a piece of pragmatism that enabled him to obtain a workable theory. Unfortunately, the physicists who came after him were so in thrall to his theory of gravity that they overlooked his reservations and became wedded to the idea of forces that acted instantaneously at a distance.

It did not matter in the slightest that Newton would have been open to Faraday's ideas, because the rest of the scientific profession believed that he would not have been. Faraday was ridiculed, and his humiliation was all the greater because he was self-taught from a humble background and knew next to no mathematics, the lingua franca of university-educated physicists.

The irony is that it was Faraday's lack of mathematical knowledge that freed him from the straitjacket of Newtonian – or, at least, supposed Newtonian – thinking and enabled him to 'see' the electric and magnetic fields that pervade space, and with the intuition he gained from this worldview to design experiments that no one else would have thought of.

Maxwell, pretty much alone among nineteenth-century mathematical physicists, recognised the importance of Faraday and his work. Like him, he had developed a fascination with the conundrum of electricity and magnetism that bordered on obsession. In February 1854, embarking on a research career after completing his graduate studies at Trinity College, Cambridge, the twenty-three-year-old Maxwell had written to the physicist William Thomson to ask his advice on what he should read in order to get his head around the bewildering array of electrical and magnetic phenomena.

Thomson, who would later become Lord Kelvin, was beginning his involvement in the ambitious scheme to lay a telegraph cable under the Atlantic between Britain and America – the Apollo Program of its day – but nevertheless found time to recommend Faraday's *Experimental Researches in Electricity*. The three-volume treatise was a masterful summary of everything that was known about the subject, much of which had been discovered by Faraday himself. In poring over its clear-cut descriptions of electrical and magnetic phenomena, Maxwell felt he was seeing into the mind of the man who had made them. Faraday was an experimenter with a crystal-clear vision who accepted nothing until he could demonstrate it himself. Maxwell was so impressed that he decided not to read any work on electricity by those who approached the subject through an analysis of forces acting at a distance until he was utterly familiar with Faraday's work.

Maxwell was particularly taken by Faraday's idea of electric and magnetic fields. In one simple demonstration Faraday had sprinkled iron filings around a bar magnet, the pattern revealed suggesting to him that there were 'lines' of magnetic force in the air around the magnet. When he had publicised this idea, it had caused other scientists to fall off their chairs with laughter, but by repeating the simple experiment Maxwell could see the truth in what Faraday had claimed.

The challenge was clear to Maxwell: to find a way of expressing Faraday's visual ideas in the language of mathematics. As a first step, he set out to concoct a 'toy model' that mimicked Faraday's results and with which he could make sense of them. It was not an easy task. He began with the idea that the magnetic and electric fields behaved like a fluid, governed by the mathematical laws of fluid flow and with the speed and direction of the flow at any point representing the density and direction of the lines of force. In February 1857, with some trepidation, he sent Faraday a preliminary paper on his progress entitled 'On Faraday's Lines of Force'. Although he had a strong feeling that Faraday was a kindred soul, he could not be sure that the older man would feel the same way about him.

He need not have worried. For Faraday, who had been humiliated by his scientific peers, reading the letter from a Cambridge-educated physicist who took his work seriously was one of the great moments of his life. He wrote back to Maxwell, 'I was at first almost frightened when I saw the mathematical force made to bear upon the subject, and then wondered to see that the subject stood it so well.'

Emboldened, Faraday asked Maxwell's opinion of his speculative idea that there might be gravitational lines of force as well as magnetic ones – something he knew was so outlandish that it was likely to be laughed at by other physicists. Maxwell took the idea seriously and sent a long and thoughtful reply, to which Faraday responded, 'Your letter is the first intercommunication on the subject with one of your mode and habit of thinking. It will do me much good, and I shall read and meditate again and again . . . I hang on to your words because they are to me weighty and . . . give me a great comfort.'

The four decades between Faraday and Maxwell precluded them from ever becoming the closest of friends, but they revered each other and shared a powerful bond: both had dared to challenge the scientific establishment, and neither could have achieved the fame he did without the other. Like Faraday, Maxwell knew what it felt like to be humiliated. His mother had died when he was only eight and he had been brought up in isolation by his father at Glenlair. On arrival for his first day at the prestigious Edinburgh Academy, the other pupils made fun of his country bumpkin accent, his social awkwardness and his homemade shoes and tunic, and christened him 'Dafty'.

Maxwell struggled for many years to explain Faraday's results. Although he started out with the idea that magnetic and electric fields behave like a fluid, he later devised a superior model. It addressed one of the most curious aspects of magnetism, which flew in the face of the Newtonians who believed a force of any kind between two bodies always acted along the line joining them. Magnetic force, as Ørsted had discovered, was circular. His compass needle, suspended beside a vertical current-carrying wire, pointed not at the wire but at right angles to it, and it continued to do so if the compass was moved around the wire. The magnetic force seemed to swirl around the wire like an invisible tornado. In fact, it was precisely this tornado that Faraday had exploited in his creation of the world's first electric motor.

In his new 'toy' model, Maxwell imagined that all space, whether empty or occupied by matter, was packed with tiny toothed cogs that were able to spin. A cog in direct contact with a magnet rotated, which turned the cog next to it, which turned the next cog, and so on. In this way, a circular force was communicated through space from a magnet to a piece of metal in its vicinity.

But invisible cogs were only the starting point for Maxwell's mechanical model. He also envisioned tiny beads that could move like ball bearings along the channels between the cogs and which represented electric currents. He continually tweaked his model to mimic more features of the real world. For instance, in an attempt to reproduce the fact that the magnetic strength of a material depends on the material, he made the ease with which the cogs inside matter turn depend on the type of matter they occupied. Finally, he made the cogs springy so that they could transmit internal forces across their bodies without losing energy. He made this last change at Glenlair in the summer before he and Katherine returned to London, and the moment he made it, he realised something hugely significant: the medium of cogs and beads he had concocted had exactly the properties necessary for the propagation of a wave.

In the case of a wave on a pond, a disturbance caused by a raindrop creates a temporary hummock of water. The existence of a restoring force – gravity – causes the hummock to collapse back down to the level of the pond. But because the water has mass, or inertia, it overshoots, so the hummock becomes a depression and the whole process repeats. But water is a continuous medium, so it does not simply oscillate up and down at one location. The disturbance is communicated to the next body of water, though with a delay, which in turn is communicated to the next body, with a further delay. In this way, a wave-like disturbance propagates outwards in concentric circles across the face of the pond.

Maxwell's medium of cogs and beads exhibited both inertia and a restoring force. Consequently, if it were jiggled, a ripple-like disturbance would propagate through it just like a wave on a pond. There was one proviso: if the medium were conducting, a wave could not be sustained for any distance because the currents it generated would quickly sap the wave of energy. Instead, a wave could be supported only in a non-conducting medium in which only the most fleeting of currents could be made to flow.* Such 'dielectric' materials included water, air and the vacuum of empty space.

Maxwell realised that such a wave would consist of an electric field oscillating at right angles to a magnetic field, with both perpendicular to its direction of travel. As the electric field decayed in strength, the change automatically generated a magnetic field. And as the magnetic field decayed, the change automatically generated an electric field. The process would happen over and over again, and once set in motion would continue forever in a self-sustaining wave of electricity and magnetism.

According to Maxwell's theory, the velocity of such an 'electromagnetic wave' depended on two parameters: the magnetic 'permeability' of the medium and its electrical 'permittivity'. The first was a measure of how well a medium boosted a magnetic field – its restoring force – and the second a measure of how much it hindered an electric field – its inertia. Maxwell knew that both quantities had been measured experimentally for a vacuum, but stuck at Glenlair over the summer he did not

---

* A dielectric material consists of molecules which have a net positive charge on one side and a net negative charge on the other. In the presence of an electric force field – which, by convention, points in the direction in which positive charges move – the molecules line up along the direction of the field. The electric field of such 'polarised' molecules always acts to oppose and reduce the applied electric field. Maxwell christened the brief current that flows as electric charge is polarised, or 'displaced', a 'displacement current'.

have the reference book that contained the relevant results. The book was in the library at King's College, which was why on this morning in October 1862 he had not waited for his cook to serve breakfast before running to catch the horse-drawn bus from Kensington High Street.

The London traffic had been terrible. It was the reason why a revolutionary underground transport system – the Metropolitan Railway – was being built between Paddington and Farringdon. Maxwell was not sure what he thought of the smoke and soot of steam trains operating below ground, but London was a city the like of which had never existed before and the underground railway was not the only massive engineering project underway in the metropolis: Joseph Bazalgette, chief engineer of London's Metropolitan Board of Works, was building a gargantuan system of underground sewers.

Finally, the bus reached its destination and Maxwell disembarked near Waterloo Bridge. Dodging the pedestrians on the Strand, he hurried past Somerset House and arrived at King's College. In the library, he quickly identified the reference book and found the data he needed from the experiments of Wilhelm Weber and Rudolf Kohlrausch. Plugging the numbers into his theory, he came up with a velocity for an electromagnetic wave in a vacuum. It was 193,088 miles per second.

Laboratory measurements made by the French physicist Hippolyte Fizeau in the late 1840s had given a figure for the speed of light as 193,118 miles per second; it was too close to be a coincidence. So not only was there a connection between electricity and magnetism, there was also a connection between electricity, magnetism and light! It was an extraordinary discovery that Maxwell had not foreseen when he had embarked on his work, but, incredibly, his calculations proved that light was a ripple in the electric and magnetic fields – a wave of electromagnetism.

One other person in the world had in fact guessed that there was a connection between electricity, magnetism and light: Faraday. In late September 1845, he had passed light from an oil lamp through a piece of lead borosilicate glass which he had placed between the north and south poles of a powerful electromagnet. When he turned on the power, he immediately observed a change in the light's 'polarisation'.* 'I have succeeded', he wrote jubilantly in his notebook, 'in magnetising a ray of light.'

'Faraday rotation' was incontrovertible evidence that light responded to magnetism, which suggested that light itself was in some way magnetic. And because magnetism was connected to electricity, it made sense that light must also be in some way electric. 'I happen to have discovered a direct relation between magnetism and light, also electricity and light, and the field it opens is so large and I think rich,' Faraday wrote prophetically.[9]

Alone in Faraday's basement laboratory, Maxwell smiled to himself as he imagined his fellow scientist's reaction to the news that he had proved the connection. To reach the proof he had stood on the shoulders of giants, and none towered higher than Faraday. Back on the street, he hardly noticed the crowds

---

* When, eventually, Maxwell discovered that light was an electromagnetic wave, with an electric field oscillating in strength at right angles to an oscillating magnetic field and with both at right angles to the wave's direction of motion, it become clear what 'polarisation' was. The electric (and hence the magnetic) field was free to vary in any plane whatsoever. This was the case in normal light, which turned out to be a mix of waves in which the electric field varied in all possible planes. By contrast, polarised light consisted of waves in which the electric field oscillated in a single plane – the 'plane of polarisation'.

on Piccadilly. As he passed Green Park, he thought about the implications of his discovery. He entered Hyde Park and headed towards the Serpentine. He had promised Katherine he would be home in time to go to the stables in Bathurst Mews. They rode most afternoons, he on a hired horse, she on her bay pony Charlie, which had made the long train journey down from Glenlair. The plan was to circle Kensington Gardens and Hyde Park; it was not a patch on their favourite ride from Glenlair to Old Bridge of Urr, but it was the best they could do in smoky central London.

He owed so much to Katherine. Although he had nursed her through much ill health, she in turn had nursed him through smallpox, which had almost killed him shortly before their move to London. She was his soulmate and scientific helper. Together, they carried out experiments in the attic of their London house, the eight-foot-long, coffin-shaped light box with which they 'painted' with sunlight horrifying their neighbours and giving them the reputation of mad eccentrics. For the thirty-two-year-old Maxwell, the sojourn in London was proving to be the most productive episode of his career.

Maxwell hurried along the footpath beside the enormous expanse of the Serpentine, created in the 1720s by King George II as a memorial to his beloved wife, Queen Caroline. To the south of the kilometre-long lake lay the site of the 1851 Great Exhibition, one of the wonders of the century. Among the visitors to the great glass-and-iron pavilion, so enormous it had enclosed some of the park's tallest trees, had been Charles Darwin, Charlotte Brontë, Charles Dickens and Alfred Tennyson. It had been disassembled bit by bit and reassembled at the Penge Place Estate in Sydenham, South London. To the southwest of its former site was 'Albertopolis', the district nicknamed in honour of the royal consort Prince Albert, who had died the previous December and whose plan it was that the Great Exhibition would leave a lasting cultural legacy in the form of museums and institutions. Maxwell had on numerous occasions visited the newly opened South Kensington Museum.[10]

A ferry was chugging across the Serpentine; swans, ducks and seagulls bobbed around it, but Maxwell paid them no attention. He was captivated instead by a rapidly fading rainbow in the sky. Ever since Piccadilly, a single thought had occupied him: his cog-and-bead model set no restriction on how fast or how slow the electromagnetic field might be jiggled, which could mean only that the colours of the rainbow represented a tiny range of possible frequencies. Beyond this visible 'spectrum', stretching away in both directions, there must exist undulations of the electromagnetic fields that were both more sluggish and more rapid than those of visible light. By convention, the rainbow contained seven colours, but in addition to these, he now realised, there must be other 'colours', invisible to the naked eye. Millions upon millions of them. It was an extraordinary, mind-expanding thought.

For a moment, standing on the path by the Serpentine amid squabbling seagulls, he was overwhelmed by a Faraday-esque vision of reality. All about him, stretching to the very fringes of the known universe, was the electromagnetic field, like a vast invisible ocean of energy in constant upheaval, its multitudinous vibrations filling the air all around him. And he was the first person in the history of the human race to realise this.

As the English biologist Francis Crick would one day observe, 'It is not easy to convey, unless one has experienced it, the dramatic feeling of sudden enlightenment that floods the mind when the right idea finally clicks into place. One immediately sees how many previously puzzling facts are neatly explained by the new hypothesis. One could kick oneself for not having the idea earlier, it now seems so obvious. Yet before, everything was in a fog.'[11]

Maxwell's mind was racing. Might it be possible to artificially vibrate the electromagnetic fields? Was it conceivable that, by means of some yet-to-be-invented technology, invisible electromagnetic waves might be created? He could see no reason why not. But it was now late afternoon and he could not afford to daydream any longer. Quickening his pace, he hurried along the bank of the Serpentine and crossed the road into Kensington Gardens. Ahead of him, in the vestibule of 8 Palace Gardens,[12] Katherine would already be dressed for her ride and waiting for him impatiently.

Karlsruhe, 12 December 1887

Heinrich Hertz knew something was leaping across the space between his transmitter and his receiver. According to Maxwell's theory, if electromagnetic waves were spreading outwards from the stuttering spark of his transmitter like a disturbance from a stone tossed into a pond, they should induce an electric current in the conducting loop of his receiver, which in turn should cause a fresh spark to jump across the gap in the loop. He could not yet be absolutely sure that was happening, but he had an idea.

It was not quick to implement; it took almost a month and the help of his assistant, Julius Amman. But now, fastened securely to the sandstone front wall of the laboratory, between its two doors, was a large sheet of conducting zinc, four metres high and two metres wide. Hertz's idea was to transmit a signal towards the zinc wall and attempt to pick up a reflection with his receiver.

It was an old idea. If a wave is reflected and propagates back through itself, the outgoing and incoming waves 'interfere' with each other. Where the peaks of one coincide with the peaks of the other, the two waves reinforce; and where the peaks of one coincide with the troughs of the other, they cancel each other out. The result is a wave that exhibits places where its amplitude is permanently large, alternating with places where it is zero. Such a 'standing wave' – most easily seen on a vigorously shaken washing line – appears frozen in space.

Hertz moved his receiver slowly towards the wall, which was twelve metres from his transmitter, and as he did so, he was amazed. The spark grew and disappeared, every three metres; it was the unmistakable signature of a standing wave, and exactly what he had expected. He and Amman had engineered the transmitter so that the stuttering spark in the gap caused an electric current to slosh back and forth along the three-metre conductor. The electric field associated with that current, changing fifty million times a second, radiated an electromagnetic wave with a three-metre separation between its peaks and troughs.

There was absolutely no doubt about it. Hertz had generated and detected Maxwell's invisible electromagnetic waves. They had a wavelength of six metres – the distance over which they repeated their up-and-down motion. The world would never be the same again.

o o o

Maxwell never had the satisfaction of seeing his prediction confirmed. He died tragically young at forty-seven of stomach cancer, which had killed his mother at the same age, after an excruciating operation without anaesthetic. But before he died, he advanced his theory of electromagnetism by one more critical step.

Most other scientists had been utterly baffled by his intricate mechanical model with cogs and beads, though Maxwell never expected anyone to take it seriously – to him it had only ever been a model of nature, not the way nature actually was. And in 1873, he knocked away the theoretical scaffolding and expressed his theory in nothing more than mathematical equations that described the behaviour of the electric and magnetic fields.

Maxwell's four equations of electromagnetism are so famous that today they are even emblazoned on T-shirts, often accompanied by the slogan 'Let there be light!' But Maxwell actually formulated a total of twenty equations to describe electricity and magnetism, and he wrote them not in terms of electric and magnetic fields but magnetic and electric 'potentials'. It was the English electrical engineer and physicist Oliver Heaviside who, in 1885, reduced them to the condensed form that has since become synonymous with Maxwell's name. (Though ironically, it is Maxwell's original formulation which has proved most useful in the developments of twentieth-century physics.) It took only a simple manipulation of Maxwell's equations to obtain a 'wave equation' that described an electromagnetic wave.

Even as a small boy, Maxwell had demonstrated intense curiosity. He'd incessantly ask the adults around him, 'What's the go o' that?' and, when they provided answers that did not satisfy him, 'What's the particular go o' that?' With his equations of electromagnetism, he had found the 'go' of electricity and magnetism.

One can only imagine what it must have felt like to have at last conquered electricity, magnetism and light. As Einstein would one day put it on conquering space, time and gravity, 'The years of searching in the dark for the truth that one feels but cannot express, the intense desire and the alternations of confidence and misgiving until one breaks through to clarity and understanding, are only known to him who has experienced them himself.'

Maxwell's equations are remarkable in many ways. First and foremost, they mark a seismic shift in our view of the universe. Since the time of Newton, physicists had used analogies from the everyday world to model the fundamental physical world, which is what Maxwell was seeking to do with his cogs and beads. But in throwing away this scaffolding, Maxwell had understood something profound about the universe: that reality is made of things – electric and magnetic fields – with no parallel in the everyday world of familiar objects. Their essence is captured only by mathematics, the underlying language of nature. In the twentieth century, this truth would be increasingly recognised by physicists as it dawned on them that gravity is the curvature of four-dimensional space–time, and that atoms and their constituents are describable only by abstract waves of probability. 'One scientific epoch ended and another began with James Clerk Maxwell,' said Einstein.

But Maxwell's theory of electromagnetism marks not only a profound change in our view of ultimate reality; it also contains the seeds of several scientific revolutions. The fact that a furnace should in theory contain an infinite number of electromagnetic vibrations – an utterly nonsensical conclusion – caused German physicist Max Planck to propose in 1900 that there must be an energy cut-off and that electromagnetic energy comes only in discrete chunks or 'quanta', the most energetic of which are too costly to be made in a furnace. This marked the birth of 'quantum theory', the modern description of atoms and their constituents.

Maxwell's theory also contains the seed of relativity. The fact that the speed of light appears in the theory as an absolute, with no reference to the motion of its source or of anyone observing it, led Albert Einstein to propose in 1905 that the speed of light is the rock on which the universe is built, while space and time are but shifting sand. In fact, the 'special' theory of relativity reveals that space and time are aspects of the same thing: the seamless entity of 'space–time'.

Special relativity provides further insight into Maxwell's theory by resolving a paradox that is at its heart. A magnetic field arises whenever an electric charge such as an electron is moving, thus changing its electric field.* But what if you were able to shrink yourself down to the size of such a mote of matter and catch up with it? Since you would now be stationary with respect to the particle, you would see an electric field, but no magnetic field. How can a magnetic field exist for one person but not for another? There is only way out of the paradox, according to Einstein: by recognising that electric and magnetic fields, like space and time, are not fundamental things; the fundamental things are the electromagnetic field and space–time.

---

* The electron, the fundamental grain of electricity, was discovered in 'cathode rays' by the British physicist J. J. Thomson in 1895. Such rays represent electricity in its naked state, travelling through the empty space of a 'vacuum tube' rather than concealed within a conducting wire. Electrons, which orbit the 'nuclei' of atoms, explain not only the phenomenon of electricity but all of chemistry, which is nothing more than a game of musical chairs in which electrons rearrange themselves within atoms.

How much electric field, magnetic field, space and time you see individually depends on how fast you are moving.

Arguably the most important aspect of Maxwell's theory of electromagnetism, however, is the concept of the field – invented by Faraday but given mathematical expression by Maxwell.

Faraday recognised that the electromagnetic field was a new kind of entity which differed from matter and could transmit effects from place to place. He intuited that electricity and magnetism are best understood via the field rather than via charged bodies and currents. When a current flows in a wire, the most important aspect of the phenomenon is not the current itself but the fields of electric and magnetic force that extend through space in its vicinity. This elevation of the field to a position of pre-eminence was Faraday's greatest and most prescient achievement. He had anticipated the future of physics.

In the modern view, it is the fields – not just the electromagnetic field but the electron field, the up-quark field, the Higgs field, and so on – that are the ultimate entities from which the universe is constructed. In this picture, the fundamental forces, including the electromagnetic force, exist for no other reason than to enforce something known as 'local gauge invariance', which I will explain in more detail in Chapter 8. It is remarkable that the electrical and magnetic phenomena discovered during centuries of experimenting and theorising turn out to be nothing more than a consequence of this incredibly simple and universal principle.

Maxwell's uniting of nature's electric and magnetic forces in his theory of electromagnetism has been called the second great unification in physics, after Newton's unification of the laws of heaven and Earth in his universal theory of gravity.

Maxwell is now widely considered to be the most important physicist between the time of Newton and Einstein. Those who taunted him at school in Edinburgh, and whose own names nobody now remembers, came to learn their mistake during their lifetimes.

Maxwell did not live to see the prediction of this theory of electromagnetism fulfilled in the creation of Hertzian, or radio, waves. But although he died tragically young, he did not die as young as Hertz, who contracted a rare disease called Wegener's granulomatosis, in which the immune system attacks the blood vessels, mainly in the ears, nose, sinuses, kidneys and lungs.[13] Despite several operations, he died of septicaemia on 1 January 1894. He was just thirty-six.

In his last letter, sent to his parents on 9 December 1893, Hertz wrote, 'If anything should really befall me, you are not to mourn; rather you must be proud a little and consider that I am among the especially elect destined to live for only a short while and yet to live enough.'[14] He had no inkling that his discovery would change the world and make possible the twentieth and twenty-first centuries. Asked about the ramifications of his discovery, he said, 'Nothing, I guess.' And, when pressed, 'It's of no use whatsoever. This is just an experiment that proves Maestro Maxwell was right.'[15]

But despite the Hertz's failure to recognise the importance of his discovery, it changed the world irrevocably. Radio, TV, Wi-Fi, microwave ovens, radar . . . the list of technologies it spawned is endless. Our ultra-connected world, which we take for granted and in which the air around us is criss-crossed by the invisible chatter of a billion voices, was born in Karlsruhe on 12 December 1887.

3

Mirror, mirror on the wall

I think that the discovery of antimatter was perhaps
the biggest jump of all the big jumps in physics
in our century.

WERNER HEISENBERG[1]

'How did you find the Dirac equation, Professor Dirac?'
'I found it beautiful.'

PAUL DIRAC[2]

Pasadena, California, 2 August 1932

It was a window onto a new world, a new universe. An odd-ball of a man six thousand miles away in England would have realised that instantly, but the young man holding up the photograph – the physicist whose persistence and hard work had obtained the image – knew only that it was extraordinary. Carl Anderson, sitting at a desk on the third floor of Caltech's Guggenheim Aeronautical Laboratory, put down the photograph and began writing the paper that would not only make his name but also make him one of the youngest people ever to win the Nobel Prize in Physics.

It had all started with Robert Millikan, the charismatic phys-icist whose relentless drive had transformed Pasadena's Throop College of Technology into the world-renowned California Institute of Technology, more commonly known as Caltech. Millikan had become intrigued by the mysterious 'radiation' that the balloon-borne experiments of the Austrian physicist

49

Victor Hess had revealed became stronger with altitude, indicating that its source was not the Earth but something in space.

At the time of Hess's discovery in 1912, the only radiation known was that which emerged from unstable, or 'radioactive', elements, such as uranium, thorium and radium. The cores, or 'nuclei', of their atoms spat out subatomic bullets, in the form of alpha particles (helium nuclei), beta particles (electrons) and gamma rays (high-energy photons). As all three types of radiation rocketed through the air, they smashed apart atoms, whose ricocheting electrons could be detected when they charged up an 'electroscope' or triggered the rattlesnake clicking of a 'Geiger counter'. Hess's 'cosmic rays' – a name coined by Millikan in 1925 – mimicked this 'ionising' effect.

At the end of 1929, with Anderson nearing the end of his PhD, Millikan asked whether he would be interested in investigating cosmic rays. It was a no-brainer for the young student; he was in awe of the Caltech president, who had won the 1923 Nobel Prize in Physics for measuring the charge on an electron.[3]

Millikan thought cosmic rays were gamma rays with enormously higher energy than any found on Earth and that electrons they collided with would ricochet like billiard balls hit by a cue ball. By measuring the energy of such 'Compton-scattered electrons', it would be possible to estimate the energy of the gamma rays.* Millikan suggested that Anderson use a cloud chamber for the task, a remarkable device invented by

---

* Arthur Compton was an American physicist who won the 1927 Nobel Prize in Physics for demonstrating that high-energy light bounces off electrons exactly as if it were made of tiny bullets. It was proof of Einstein's 1905 claim that light travels through space as a stream of particles, or 'photons'.

Charles Wilson in Cambridge in 1911 which could reveal the tracks of subatomic particles. Its principle was simple and had been copied directly from nature. When moist air rises, it cools and condenses into droplets, forming clouds. Wilson mimicked this process by filling a glass chamber with moist air and suddenly cooling it. Air naturally cools when it expands, so he was able to achieve this by pulling out a piston to increase the volume of the air.

A water droplet will form only if there is a 'seed' such as a grain of dust around which it can condense; if the water vapour is so pure that it contains no such impurities, the condensation seeds may be provided by tiny charged 'ions' created when electrons are stripped from atoms by ionising radiation.

Wilson filled a glass chamber with ultra-pure water vapour and cooled it below the temperature at which droplets would normally form. In this 'supercooled' state, the water vapour was desperate to form droplets around any ions, and would do so the instant Wilson expanded and cooled his cloud chamber.

Operating the device proved to be more of an art than a science, but by illuminating the chamber with a bright light it was possible to photograph the tiny trails of bead-like water droplets left in the wake of a passing subatomic particle. Given how mind-bogglingly tiny subatomic particles are – a million million times smaller than the smallest speck visible to the naked eye – revealing their tracks was a stunning achievement. It earned Wilson a share of the 1927 Nobel Prize in Physics.

Millikan knew that if the water droplets were spread out, leaving a thin track, the particle responsible carried a relatively small electric charge; whereas, if the water droplets were crowded together, making a thick track, the particle had a relatively large charge. The charge would help to identify a particle created by a cosmic gamma ray but would not be enough to pin down its identity definitively. Millikan therefore suggested that Anderson place his cloud chamber in a magnetic field; this would bend the path of subatomic particles, with those of low momentum being more curved than those with high momentum. (Momentum, which is the product of a body's mass and velocity, reflects the fact that a slow-moving, massive body is as hard to deflect as a fast-moving, light body.)

However, the cosmic rays and their subatomic debris were extremely penetrating – they were capable of smashing their way through a thick sheet of dense lead – which indicated that they had enormous energy and were travelling extremely fast. Such a fast-moving particle would spend very little time traversing the cloud chamber, which meant the magnetic field would have little opportunity to bend its trajectory. The only way to create a measurable deflection was to use the strongest possible magnetic field.

The experiment was a major challenge and it took a full year to assemble the apparatus in the optical shop of the Robinson Laboratory of Astrophysics, which had been set up to build a world-beating five-metre telescope for Mount Palomar Observatory near San Diego.[4] The Great Depression, which had been triggered by the Wall Street Crash of 29 October 1929, was in full swing: money was tight, and Anderson had to salvage material for his experiment from local scrapyards. Fortunately, he had a long history of improvising with discarded equipment, having powered electrical experiments with used automobile batteries he had cadged from garages while a high-school student in Los Angeles.[5]

Anderson's cloud chamber was the shape of a very shallow biscuit tin, three centimetres deep and seventeen centimetres in diameter. It was arranged with its long dimension vertical, and embedded in the coils of a solenoid – a tight coil of copper wire carrying an electric current. The bigger the current, the stronger the magnetic field, and the biggest current available at Caltech was produced by the 425-kilowatt generator that powered the wind tunnel in the Guggenheim lab. This was why Anderson had installed his apparatus in the aeronautics building.

Large electric currents generate large amounts of heat, which was another major problem for Anderson. In order to stop his apparatus melting, he had to pump cooling water through pipes which spiralled around his solenoid. Although the cloud chamber at the heart of the experiment had a diameter barely greater than a tea plate, the entire apparatus ended up weighing close to two tonnes.

When operating, it was a fearsome sight. Tap water was pumped through at a rate of forty gallons a minute. Heated to almost boiling point by the current surging through the sole-noid, it was piped out of the lab, down the outside wall of the Guggenheim lab and to a drain on the far side of neighbouring California Street (now California Boulevard). Anderson had no choice but to work at night because the Guggenheim's genera-tor was needed to power the wind tunnel during the day. Local Pasadena residents were alarmed at the sight of steam billowing upwards in the dark between the spindly palm trees on Arden Road, and it required all Millikan's charm and powers of diplo-macy to reassure them that their lives were not at risk.

Even more dramatic were the supernova-bright flashes of light that came intermittently from the third-floor window of the Guggenheim lab.[6] Anderson used a powerful arc lamp to illuminate particle tracks and a camera to record them. For those strolling down Olive Walk after a pleasant dinner at

Caltech's new Athenaeum club, it seemed as if Frankenstein's monster was being brought to life. Had anyone climbed the stairs to the third floor, the sight of Anderson, in white coat and welding glasses, crouched beside the magnetic coils of his apparatus, would have done little to put their mind at rest.

The cosmic ray experiment had a soul-destroyingly low success rate. The arrival of particles from space was unpredictable, so it was never possible to know when one was flying through the cloud chamber. Anderson's only option was to activate the piston at random moments, illuminate its interior with a flash of light, take a photograph and hope for the best. Not surprisingly, the majority of the photographs were blank. In fact, after a year of operation, during which Anderson took some 1,300 photographs, only fifteen turned out to contain anything of interest, a success rate of barely 1 per cent. Thomas Edison, who famously called genius '1 per cent inspiration and 99 per cent perspiration', had hit the nail on the head.

In a handful of the photographs, however, it was possible to see the track of a lightweight, low-momentum particle, curving tightly in the enormous 15,000-gauss magnetic field. It could only be an electron, created in the cloud chamber by the collision of a cosmic ray with an atomic nucleus. But Anderson noticed something curious: in addition to showing an electron corkscrewing one way in the magnetic field, pretty much every photograph showed the track of a particle corkscrewing in the opposite direction. That could mean only that it was a positively charged particle, since a magnetic field bends positively charged particles the opposite way to negatively charged ones (and electrons always carry a negative charge). The thickness of the tracks revealed that the charge carried by the particles was of exactly the same magnitude as that of an electron.

The idea that a positively charged electron existed was too ridiculous to contemplate. Something must have gone awry with Anderson's experiment.

Millikan, though a meticulous experimenter, was prone to eccentricity and poor judgement. He had the half-baked idea that cosmic rays were gamma rays generated in deep space by the birth of atoms.[7] If he was right, the high-energy photon of a gamma ray should simply give a single electron an almighty kick. However, this was not at all compatible with Anderson's photographs, which showed the creation of roughly equal numbers of negative and positive particles, often speeding away from a common point.

The only positively charged particle known at the time was the proton, but that was about two thousand times more massive than an electron and the tight curvature of Anderson's tracks indicated that the mystery particle was considerably lighter than that.

On inspecting the tracks, Millikan asked his assistant whether a gamma ray could have interacted with the cloud chamber's glass base, sending an electron speeding back upwards? His reasoning was that the magnetic field would bend the path of an electron travelling upwards in the same way as it would a positively charged particle travelling downwards.

There was an easy way to find out. Anderson inserted a horizontal lead sheet across the middle of his cloud chamber. Inevitably, a particle ploughing through it would lose speed, causing it to spend more time in the magnetic field so that its track would become more curved. The part of the track that was more tightly curved showed the particle at a later time, and therefore indicated its direction of travel.

As Anderson modified his experiment, the 1932 Olympics were in full swing in nearby Los Angeles. Due to the Great

Depression, only half as many athletes could afford to attend as had competed in the 1928 Olympics in Amsterdam. As part of the games' money-saving effort, Pasadena's Rose Bowl Stadium had been converted into a velodrome; it was not far from Caltech and, if Anderson had a spare moment, he intended to watch some of the cycling. Despite the Olympic excitement, however, on the Caltech campus, in the shadow of the San Gabriel Mountains, all was quiet. The students were away and many of the faculty had gone on vacation to escape the sweltering July heat.

Once again, Anderson knuckled down and took lots of photographs, most of which turned out to be useless. But on 2 August 1932, he obtained a heart-stopping image. He was staring at it now as he penned the introduction to his paper.

Crossing the middle of the photograph was a thick black horizontal line, the shadow of the lead plate. Above the line, the track of the particle – no thicker than a human hair – was more tightly curved than below it, confirming that the particle was indeed travelling upwards not downwards – a rare event that historians of science would debate much in the future. But it was not the fact that the particle was travelling upwards that made the picture heart-stopping; it was the nature of the track itself. It was curving the wrong way.

Millikan dismissed the track as a freak, and Anderson himself had nagging doubts. But taken at face value, the track could be interpreted only as the trajectory of a light particle like an electron, but carrying a positive rather than a negative electric charge. For a moment, Anderson hesitated over the page he was writing. Then, for the first time, he wrote down the word he had coined for the new particle. He called it a 'positron'.

Anderson's paper, which he intended to submit to the American journal *Science*, was entitled 'The Apparent Existence of Easily Deflectable Positives'. 'It seems necessary to call upon a positively charged particle having a mass comparable with an electron,' he wrote. It was a controversial statement. But what could he do? He had no choice but to accept what his experiment was telling him. There on his photograph, as clear as day, was the unmistakable signature of a positively charged electron.

In 1932, only three fundamental constituents of matter were known: the electron, the proton and the neutron, whose discovery had been announced in February that year by James Chadwick at the University of Cambridge.[8] Between them, the three particles provided the basic building blocks of an atom, in which electrons circled a tight ball of protons and neutrons like planets around the Sun. It was a neat and appealing picture of the structure of matter, and the last thing anyone wanted was for another particle to mess things up. Nobody needed a positive electron. Nature had no place for it. Or did it?

Cambridge, Late November 1927

When he first wrote down the equation that described the electron, Paul Dirac was stunned and awed by its beauty,[9] but he was also terrified. He felt like a tightrope walker who had achieved a miraculous balancing feat but might be sent plunging to his death by the slightest puff of air. His equation was a piece of magic whose beauty was the mysterious hallmark of something that was right, but what if he was deluding himself? What if there was an ugly fact out there waiting to kill it stone dead?[10] He had to take deep breaths to ward off a panic attack.

Tall, gangly and reminiscent of a stick insect, Dirac was the strangest of strange men. For six days each week he worked hard, and then on Sundays he cleared his head with a long walk in the countryside outside Cambridge, where he would climb trees while dressed in a suit and tie. Literal to the point of obtuseness, Dirac could have out-Spocked Mr Spock. When a student in one of his classes put up their hand and said, 'Professor Dirac, I don't understand the equation on the blackboard,' he replied, 'That's a comment not a question,' and looked into the middle distance.[11] One of Dirac's friends, the Russian physicist Peter Kapitza, tried to get Dirac interested in Russian literature and gave him a copy of *Crime and Punishment*. When he had finished it, Kapitza was eager to know what he had thought. Dirac's one and only comment was, 'In one of the chapters the author has the sun rising twice on the same day.'

Dirac could spend hours in the company of others without feeling the slightest obligation to utter a single word, and if he did decide to speak, his conversation would often be limited to 'yes' or 'no'. But although Dirac seemed baffled by the world of everyday social interactions – though not by the cartoon world of Mickey Mouse, which was one of his peculiar obsessions – he was not baffled by the abstract realm of fundamental physics. He was a high priest of quantum theory and Einstein's theory of relativity.

By the autumn of 1927, the problem of how to unite quantum theory and relativity in a description of the electron had been occupying Dirac for many months. It had been occupying many physicists, in fact. It was, after all, the obvious problem.

Quantum theory was the description of the microscopic realm of atoms and their constituents. It was fantastically successful and predicted the results of many experiments extremely accurately. But in addition to its success, it provided a window onto a bizarre, counter-intuitive, Alice-in-Wonderland world that lurked just beneath the skin of reality. It was a place where a single atom could be in many places at once; where things happen for no reason whatsoever; and where two atoms could influence each other instantaneously, even if on opposite sides of the universe.

Much of this quantum weirdness arose from a single extraordinary observation: the fundamental building blocks of matter – electrons, protons and photons – could behave both as localised, bullet-like particles and as spread-out waves like ripples on a pond. They are like nothing in the familiar everyday world. The first hint of this microscopic madness had come in 1900, but it had not been until the mid-1920s that physicists formulated a fundamental theory from which it was possible to make precise predictions of the behaviour of the atomic world.

The high point of quantum physics was the Schrödinger equation, devised by Austrian physicist Erwin Schrödinger in 1925. It melded together the particle and wave behaviours, and described how 'quantum waves' spread through space, their 'amplitude' (strictly speaking, their 'amplitude-squared') at any location determining the 'probability' of finding a particle there.[12] The Schrödinger equation, however, had a problem: it was not compatible with the other great development of early-twentieth-century physics, relativity.

Einstein's special theory of relativity, published in 1905, recognised that the speed of light is the rock on which the universe is founded and that space and time are but shifting sands. In fact, at speeds approaching that of light, space and time blur into each other, revealing that they are aspects of the same thing: space–time. If it were possible for someone to fly past you at close to the speed of light, their time would slow so that they would appear to be moving through treacle and their space would shrink in the direction of their motion so that they appeared flattened like a pancake.*

These counter-intuitive effects are noticeable only at speeds approaching that of light, which at a million times faster than a passenger jet is way beyond anything we experience in the everyday world. The Schrödinger equation is therefore perfectly adequate for describing a hydrogen atom, in which an electron orbits the solitary proton in the nucleus at less than 1 per cent of the speed of light. However, the electric force which binds electrons to an atomic nucleus gets stronger the more protons there are in a nucleus. In the heaviest atoms, such as uranium, the force can whirl electrons around at speeds approaching that of light.† The Schrödinger equation is inadequate for describing such particles; an equation that was compatible with special relativity was needed, and this is what Dirac had been pursuing.

The challenge was to generalise Schrödinger's equation for an electron – to find an overarching formula of which Schrödinger's relation would turn out to be merely a special case when the speed was much less than that of light. There is no recipe for generalising an equation in physics – it involves intuition, guesswork and courageous leaps of faith. It is like

* Actually, this is not quite true. Although special relativity predicts that someone moving relative to you should appear to shrink in the direction of their motion, this is not what you would see because of another effect at play. Light from more distant parts of the person takes longer to reach you than from closer parts, which causes them to appear to rotate. So if their face is pointing towards you, you will see some of the back of their head. This peculiar effect is known as 'relativistic aberration' or 'relativistic beaming'.
† The electromagnetic force between a proton and an electron in a hydrogen atom is ten thousand billion billion billion billion times stronger than the gravitational force between them.

being in an unknown land at midnight without a torch or map and trying to guess the topology of the landscape. There were clues, however. Since Einstein had shown that space and time were aspects of the same thing, Dirac knew that the equation he was seeking must treat space and time equally. It must also incorporate another key aspect of special relativity: the idea that mass is a form of energy.

A cornerstone of Einstein's theory is that light is uncatchable: its velocity, for some unknown reason, plays the role of infinite speed in our universe. The only way light can be uncatchable is if a material body resists being pushed to the speed of light; such resistance, or 'inertia', is the very definition of mass. A body must therefore become more massive as it approaches the speed of light. Since the only thing that obviously increases as the body accelerates is its 'energy of motion', the unavoidable conclusion is that energy of motion has mass. In fact, as Einstein realised, all forms of energy, and not just energy of motion, have an equivalent mass.

But just as energy has mass, mass has energy. Einstein's most remarkable discovery was arguably that energy is locked away inside matter, even when it is at rest. Mass-energy is the most compact and concentrated form of energy, with the amount available stupendously large and given by the most famous formula in science: $E = mc^2$, where c is the speed of light.

The expectation might be that the total energy of a particle travelling at an appreciable fraction of the speed of light is equal to its rest energy plus its energy of motion. However, according to Einstein, it is more complicated than that: it turns out that the square of a particle's total energy is equal to the square of its rest energy plus the square of its energy of motion. It is therefore necessary to take the square root of this expression to obtain the energy. Immediately, however, this creates a problem. Just as the square root of nine can be both minus three and three, the square root of the expression for the 'relativistic' energy can be negative. This was a nonsensical result that Dirac wanted to avoid at all cost, so he set out to find an equation that directly yielded the energy of a particle and not its energy squared.

The technical question was: How could he obtain an expression for the energy of an electron as a sum of a multiple of its rest energy and another multiple of its energy of motion? This task turned out to be impossible if the two multiples are numbers. For anyone else this would have been an impasse, but Dirac's genius was to realise that it was possible to obtain an expression for the energy if each of the multiples, rather than being a simple number, was a 'two-dimensional number' – a table of values with two rows and two columns.

Mathematicians have special rules for adding and multiplying together such 'matrices'. A key property is that multiplying matrix A by matrix B does not necessarily give the same result as multiplying matrix B by matrix A, which is not an uncommon property of 'operations' in the everyday world. Take a die. If it is rotated ninety degrees clockwise about a vertical axis and then ninety degrees top to bottom about a horizontal axis, its final orientation is not the same as if it is turned ninety degrees top to bottom, then ninety degrees clockwise.[13] Since a die keeps track of what happens when it is rotated and the matrices Dirac required to describe a relativistic electron did the same, it provided a hint that an electron can in some sense rotate; that is, it has 'spin'.

Such a property had been revealed in experiments and had completely baffled theorists. Electrons flying through a magnetic field were deflected in two distinct ways, as if they were miniature magnets that could point either in the direction of the field and be deflected one way or in the opposite direction and be deflected the other way.[14] Magnetic fields are generated by electric currents, which are simply electric charges in motion. And the only way the charge on an elementary particle like an electron could be moving is if the electron is spinning.

However, calculations showed that, for an electron to generate the strength of magnetism revealed in experiments, it would have to be spinning faster than the speed of light, which, according to Einstein, was impossible. Physicists were forced to accept that an electron behaves as if it is spinning, even though it is not. Its intrinsic 'quantum spin' is a property with no analogue in the everyday world, but nevertheless it has real effects. If a large number of electrons ran into you, they would impart their intrinsic spin to you and you would find yourself spinning like a pirouetting ice skater.

The fact that the matrices Dirac used to describe the electron used two columns of paired numbers implied a 'two-ness' to the spin, which was what had been observed. Although spin appeared nowhere in the Schrödinger equation, it emerged quite naturally from the mathematics of Dirac's matrices.

Dirac worked in a study at St John's College in Cambridge that had no pictures on the wall nor any ornaments or other frivolity; were it not for an ancient settee against one wall, it would have been indistinguishable from an empty classroom. He worked best in the early mornings, seated at a simple folding desk, with his head down, scribbling on scraps of paper with a pencil and occasionally pausing to rub out an error or to check something in one of his handful of reference books. The silence of his study was interrupted only by the creaking of his door as his man-servant, or 'gyp', crept in to add coal to his fire or to bring him tea and biscuits.

By late November, Dirac had tried and discarded many mathematical formulations; it was then that he conjured a description of the electron which simultaneously respected the constraints of both theories, squaring what had seemed an impossible circle.[15] He could hardly believe that he had finally found what he had been looking for. What convinced him was that the formula he had concocted appeared to have something of the divine about it. It was economical, elegant and beautiful. He, a human being, had invented it, but it could have been a thought from the Creator that had wafted down from heaven and landed on his page.

Dirac's equation described not only a particle with the mass of an electron but one with exactly the same spin and magnetic field as had been found in experiments. The definitive test, however, would be to apply it to nature's simplest atom, hydrogen. The 'energy levels' of its single electron had been pinned down by experiments to a high degree of precision, though Dirac's fear of falling from a great height was so great that he could not bring himself to make predictions with his equation. Instead, he carried out only an approximate calculation; to his relief, his predictions chimed with reality, but he dared go no further.

For almost a month, Dirac told nobody about his discovery; he broke his silence only on the eve of leaving Cambridge for his parents' home in Bristol for the Christmas vacation, when he bumped into Charles Galton Darwin, grandson of the biologist and a leading theoretical physicist. Darwin was deeply impressed by what Dirac told him, and on Boxing Day wrote to the Danish quantum physicist Niels Bohr: '[Dirac] has now got a completely new system of equations which does the spin right in all cases and seems to be "the thing".'

Dirac submitted a paper to the Royal Society on 1 January 1928, and it appeared in print a month later.[16] 'The Quantum Theory of the Electron' caused a sensation. According to the American physicist John Van Vleck, Dirac's explanation of an electron's spin was comparable to 'a magician's extractions of rabbits from a silk hat'.

The successes of Dirac's equation, however, came at a cost. At the outset, he had attempted to exclude the possibility of negative-energy electrons, but he had failed miserably. His beautiful equation contained not one but two sets of 'two-by-two' matrices – one representing positive-energy electrons and the other negative-energy electrons.*

In 'classical', or pre-quantum, physics, it was not unusual for a theory to throw up such nonsensical 'solutions', but physicists simply dismissed them by saying that nature chooses not to implement them. In quantum theory, however, no such option is available; according to American Nobel Prize-winning physicist Murray Gell-Mann, 'Everything that is not forbidden is compulsory.' In other words, a particle such as an electron has a non-zero 'probability' of making a 'transition' from any state to any other state – and that included Dirac's negative-energy states.

Dirac had discovered that Einstein's special theory of relativity could be satisfied by electrons only if they had spin, which was an enormous triumph, but he had also discovered it could be satisfied only if electrons were permitted to have both positive and negative energies, which was disastrous.

---

* In fact, to incorporate both in his equation, Dirac was forced to use matrices with four columns with four numbers in each, which would later become known as the 'gamma matrices'.

Physicists were amazed by the beauty of Dirac's equation and stunned by its power to predict things about the real world, but many were unsettled by its prediction of negative-energy electrons. To Werner Heisenberg, this was evidence that the equation was sick and quite possibly wrong. 'The saddest chapter of modern physics is and remains the Dirac theory,' he wrote despairingly to Wolfgang Pauli, who agreed. In Pauli's opinion, the sickness of Dirac's equation was incurable, and the agreement of its predictions with experiments was little more than a fluke.

Dirac himself did not share the misgivings of other physicists at the troubling negative-energy feature of his equation. Although his forte was the most abstract fundamental physics, he had trained as an electrical engineer and was at his core a pragmatist. If something worked – and his equation worked in predicting to unprecedented levels of precision much that had been observed in experiments – then he was sure it must contain a large amount of truth. If it failed in some respects, it might simply need some tweaking, so all he had to do was find a way to do that.

A major reason for the despair of Heisenberg was that the negative energy 'solutions' of the Dirac equation threatened the very stability of matter. In the everyday world, objects tend to reduce their 'potential energy' – that is, energy with the potential to, in physicists' jargon, do 'work'. For instance, given the chance, a ball at the top of a hill will race to the bottom, converting its potential energy into energy of motion. At the top of the hill, it is said to have 'high gravitational potential energy', and at the bottom 'low gravitational potential energy'.

The problem with Dirac's equation was that, if the negative-energy 'states' were available to electrons, there was nothing to stop them minimising their potential energy by dropping into those states. It was as inevitable as a ball rolling to the foot of a hill: matter would be unstable. Dirac's equation spelled catastrophe for the world.

Things did not look good, but in the autumn of 1928, Dirac came up with a radical idea for avoiding the disaster. Arguably, it was one of the most ridiculous ideas in the history of science.

Matter is stable, he pointed out, so all the electrons in the universe have not, by definition, dropped into the negative energy states. The obvious explanation was that his equation was wrong and that the negative-energy states did not exist, but the equation had scored so many spectacular successes that he did not want to abandon it. He therefore proposed an alternative explanation for why the electrons in matter have not dropped into the negative-energy states. There is no room for them. Why? Because those states are already filled to the brim with negative-energy electrons.*

The fact that the idea appeared nuts was not necessarily grounds to dismiss it. The key question was: Did it contradict reality? Surely we would notice if we are living in the midst of a vast sea of negative-energy electrons, but Dirac reasoned that

* The fact that each negative-energy state is filled when it contains just one electron is important. If any number of electrons could pile into a single negative-energy state, there would be no way they could ever be 'filled up' and stop normal electrons falling into them and making matter unstable. However, quantum theory permits the existence of two distinct types of particle: those with half-integer spin and those with integer spin. The minimum possible quantity, or 'quantum', of spin is half of a certain quantity. Particles with half-integer spin, known as 'fermions', have the property of being hugely antisocial and need to be one to each quantum state, while the latter, known as 'bosons', are extremely gregarious and happy to pile in together into a single state. Electrons, it turns out, are fermions.

we would not. Do we in normal circumstances notice the air around us? Do fish notice the water through which they swim?

By postulating a vast sea of negative-energy electrons to fix the difficulty with the stability of matter, Dirac was able to sweep a major problem of his equation under the carpet, but in doing so he created another headache. A vast sea of negative-energy electrons would, not surprisingly, have consequences. Occasionally, for instance, a negative-energy electron might be struck by a photon; if ejected from the sea with sufficient energy, it would become a normal positive-energy electron.

The sudden appearance in the world of an electron like a rabbit plucked out of a hat was a startling enough notion, but when Dirac followed his reasoning through to its logical conclusion, he realised something else: the ejected electron would leave behind a gaping hole in the sea of negative-energy electrons. He knew of experiments in which the inner electron in an atom was ejected by a high-energy 'X-ray' photon and left behind a similar absence; in these cases, this 'hole' behaved exactly like a positively charged electron. Dirac proposed that the hole left behind when an electron is ejected from the negative-energy sea behaves exactly like a positively charged particle. In other words, a high-energy photon would create not one particle but two: an electron plus a positively charged mirror image of an electron, in a process that would become known as 'pair production'.

Dirac was not brave enough to propose the existence of an new subatomic particle with a mass equivalent to an electron but an opposite electric charge on the basis of a mathematical formula he had conjured out of thin air, so he chose a more cautious option. At the time, the only known positively charged subatomic particle was the proton, so Dirac proposed that the positively charged mirror image of the electron was a proton. The fact that such a particle is about two thousand times the mass of an electron, marring the neat symmetry of pair production, was a detail to be sorted out later. A new fundamental particle would have been surplus to requirements and would have been strongly resisted by physicists. It was a battle Dirac did not wish to fight – so he dodged it.

According to Dirac's friend Peter Kapitza, he never seriously believed that his positively charged particle was a proton. He proposed it simply so he would not have to face other physicists mocking him with the question, 'Where is your antielectron, Professor Dirac?'

In actual fact, the proton was never a serious contender for the mirror image of the electron conjured into existence in pair production. The American physicist Robert Oppenheimer, who would one day lead the Manhattan Project to build an atomic bomb, pointed out that if a high-energy photon could create an electron and a proton, then the reverse process would also be possible, with a proton and an electron annihilating each other. This would cause matter to be dangerously unstable. Atoms would survive only as long as their protons did not run into stray electrons. At every instant, they would be prone to disappearing in a flash of gamma rays.

It was largely Oppenheimer's argument that emboldened Dirac to go public with what he already knew in his bones. In May 1931, he submitted another paper to the Royal Society. It was on a different subject entirely – a speculation on why electric charge comes in discrete chunks, or 'quanta'; however, in the paper Dirac predicted 'the existence of a new kind of particle, unknown to experimental physics, having the same mass and charge as an electron'.[17] He called it an 'antielectron' and wrote, 'We should not expect to find any of them in nature, on account of their rapid rate of recombination with electrons, but if they could be produced experimentally in high vacuum they would be quite stable and amenable to observation.'

During a lecture at Princeton University in late October 1931, Dirac went further. 'Antielectrons are not to be considered as the mathematical fiction,' he said. 'It should be possible to detect them by experimental means.'[18]

What Dirac imagined were two high-energy photons colliding and conjuring an electron and an antielectron into existence. He was not optimistic about the imminent detection of such a process, since photons of the extremely high energy required were unlikely to be available to experimenters for the foreseeable future. Dirac must have been aware that cosmic rays possessed extremely high energies – typically thousands of times higher than those of the particles spat out by the nuclei of radioactive atoms – and that they might create antielectrons when they slammed into particles in the atmosphere, yet he appeared to give them little thought. This was possibly because the experimenters he knew at Cambridge considered them insufficiently interesting to study and thought Millikan was wasting his time.

Dirac did not only predict a positively charged partner of the electron; in his Royal Society paper of May 1931, he pointed out that just as a relativistic description of the electron implied the existence of an antielectron, a relativistic description of the proton implied the existence of an antiproton. Nature must have duplicated all its fundamental particles, and there existed a mirror world of positive electrons and negative protons – a universe of 'antimatter'. 'My equation', Dirac later confessed, 'was smarter than I was.'[19]

Dirac had well and truly stuck his neck out in writing down an equation, motivated by nothing more than a desire to make quantum theory and special relativity mathematically consistent, which predicted much of what physicists observed in the world, including the existence of quantum spin. But remarkably, it also predicted that the stuff of the world that had hitherto seemed immutable – the fundamental particles of matter – could be created and destroyed at will. And if that was not shocking enough, in order for such processes to occur, there must exist a mirror universe of antimatter.

Rarely in the history of science has a single equation predicted so much novelty. 'There is something fascinating about science,' Mark Twain observed. 'One gets such wholesale returns of conjecture out of such a trifling investment of fact.'[20] Of no piece of science has that been more true than the Dirac equation.

Postulating the existence of a subatomic particle that nobody had ever seen and for which there had never been any need was controversial to say the least, but the proof of the pudding would be in the eating. For Dirac, the big question was: Did antielectrons really exist?

Pasadena, California, Autumn 1932

Carl Anderson had found no other tracks showing a positively charged particle with the mass of an electron. It was such a worry that he considered asking the journal *Science* to withdraw his paper. Had he done so, however, it would have been too late; the printing presses were already rolling.

On 1 September 1932, the paper appeared, and the reaction from other physicists was either indifference or outright disbelief. Ed McMillan, a good friend from Anderson's undergraduate days at Caltech, waved his copy of *Science* under Anderson's nose. 'What sort of nonsense is this?' he asked. Millikan, who had become convinced that there was something wrong with the cosmic ray experiment for it to yield such an incomprehensible result, was not supportive either. Anderson, his confidence undermined, wondered whether he had been a total fool and whether, at the age of twenty-seven, he had inadvertently sabotaged his scientific career.

Perhaps if Anderson had realised what he had found, it might have made a difference, but he had no idea that the particle he had detected had been predicted at a desk in Cambridge by Paul Dirac. Weirdly, he had recently been attending evening lectures given by Oppenheimer, who spent several months of each year at Caltech, and they had dealt with Dirac's hole theory at length. Anderson failed to make the connection between a Dirac hole and the peculiar particle he had discovered in his Guggenheim cloud chamber, but Oppenheimer's own blindness was arguably even more bizarre. Despite knowing about Dirac's prediction of a positively charged electron and Anderson's discovery of a positively charged electron, he unaccountably failed to put the two things together.

One of Anderson's colleagues did make the connection. Rudolph Langer, a mathematician, knew about Dirac's theory of the antielectron and had seen Anderson's photograph of the track of a lightweight, positively charged particle. Shortly after reading Anderson's paper in *Science*, he submitted a short response to the journal in which he categorically claimed that the particle Anderson had detected was Dirac's antielectron. Unfortunately, Langer was neither well known nor respected in physics circles and his paper was ignored.

It was even worse six thousand miles away in Cambridge, where nobody seemed to be aware of Anderson's experiment nor of Langer's paper in *Science*. It would take an independent experiment to wake up the physicists at the Cavendish Laboratory.

Patrick Blackett had belatedly got into cosmic ray research after Millikan had lectured at Cambridge the previous year and shown some intriguing cloud chamber photographs taken at Caltech that were, of course, Anderson's. Blackett persuaded the Cavendish Laboratory director, Ernest Rutherford – the greatest experimental physicist of the age and discoverer of the atomic nucleus – to let him get into cosmic ray research. He teamed up with the Italian physicist Giuseppe Occhialini, and the pair hit on the clever idea of observing the debris from cosmic rays by using Geiger-Müller tubes in conjunction with a cloud chamber.

A Geiger-Müller tube, or Geiger counter, consists of a gas-filled glass tube. When a particle of radiation passes through it, it knocks electrons from molecules of the gas, which are amplified by a high voltage into a measurable electric current. By putting one Geiger counter above their cloud chamber and one below it, and triggering the chamber only if both Geiger counters registered a current, Blackett and Occhialini ensured that every photograph they took contained particle tracks. Whereas Anderson's experiment had been a lottery, with the odds of detecting positrons stacked against him, this experiment was a dead cert, and the particles were photographed in large numbers.

Dirac was always hazy about how he heard about the discovery of antielectrons, but he probably heard about them from Blackett. The pictures he had obtained with Occhialini of positrons in cosmic ray showers were so sensational that they were featured on the front pages of newspapers. By mid-December 1932, there could no longer be any doubt, and Dirac confirmed that the pictures of pair production obtained at the Cavendish Laboratory were consistent with his theory. His days of panic attacks were over, and no fact, either experimental or theoretical, was going to wreck his beautiful equation. His triumph was to predict for the first time in scientific history the existence of a new fundamental particle. And he had done it using a theory he had pulled out of thin air, with pretty much no motivation from any experiment.

Blackett and Occhialini had undoubtedly provided the best evidence of the existence of the positron. In fact, Blackett had observed the positron's effects even earlier than Anderson, though he made the mistake of dismissing them as unimportant. Despite his own pivotal contribution, Blackett was always scrupulously careful to give credit to Anderson for being the first to announce the existence of the positron.[21]

In 1936, Anderson was rewarded with the Nobel Prize in Physics, sharing it with Hess, the discoverer of cosmic rays. By this time, Dirac had also been honoured, having shared the 1933 Nobel Prize with Schrödinger.

Anderson would be part of a Nobel Prize-winning dynasty of three generations of experimental physicists. Not only did his supervisor, Robert Millikan, win the prize but so too did his student, Donald Glaser, who carried off the 1950 prize for the invention of the 'bubble chamber', which revealed the tracks of subatomic particles in a similar way to the cloud chamber.

o o o

With hindsight, pair production and the existence of antimatter should have come as no surprise to anyone. It turns out that something like it is essential to unite the quantum and relativistic descriptions of the electron, or of any subatomic particle.

One of the cornerstones of physics is that energy can neither be created nor destroyed, but only transformed from one form into another. In a world ruled by special relativity, where mass itself is a form of energy, this 'law of conservation of energy' has an unavoidable consequence. The energy of motion of photons can be transformed into the mass-energy of subatomic particles – creating matter – and the mass-energy of subatomic particles can be converted into the energy of motion of photons – destroying matter.

But quantum theory applies a crucial restriction on the processes of creation and destruction: electric charge, like energy, cannot be created or destroyed. The 'law of conservation of electric charge' means that, in the creation of matter, a photon, which has no electric charge, cannot change into a subatomic particle which has an electric charge. However, a photon *can* change into two identical particles which carry opposite charges, so that their net charge is zero. Similarly, in the destruction of matter, a charged particle cannot change into a photon. This requires two identical particles with opposite charge. Thus we are led to the idea that the creation of matter must involve a photon spawning a particle and an antiparticle – pair production – and the destruction of matter must involve a particle and antiparticle spawning a photon – annihilation (yet another restriction, known as the 'law of conservation of momentum', dictates that the annihilation of matter and antimatter must result in two identical, oppositely directed photons).

'Think binary,' said the novelist John Updike. 'When matter meets antimatter, both vanish, into pure energy. But both existed; I mean, there was a condition we'll call "existence". Think of one and minus one. Together they add up to zero, nothing, nada, *niente*, right? Picture them together, then picture them separating – peeling apart . . . Now you have something, you have two somethings, where once you had nothing.'

The Dirac equation, as Updike emphasised, unveiled a world of matter and antimatter created out of absolutely nothing. There was a certain pleasing symmetry in the fact that the equation that did this had itself been conjured out of nothing by Dirac. Today it is universally admired by physicists. 'Of all the equations of physics, perhaps the most "magical" is the Dirac equation,' says American physicist and Nobel Prize winner Frank Wilczek. 'It is the most freely invented, the least conditioned by experiment, the one with the strangest and most startling consequences.'[22]

Dirac is considered pre-eminent among the magicians of science. His equation is inscribed on a square tablet commemorating the physicist on the floor of London's Westminster Abbey.

It is not just the beauty of the equation that is universally admired, but the sheer intellectual bravery of Dirac in formulating it. 'He made a breakthrough, a new method of doing physics,' said Richard Feynman, another Nobel Prize-winning physicist. 'He had the courage to simply guess at the form of an equation, the equation we now call the Dirac equation, and to try to interpret it afterwards.'[23] Feynman, a man also widely recognised as a magician of physics, was unable to go where Dirac went. 'I think equation guessing might be the best method to proceed to obtain the laws for the part of physics which is presently unknown,' he said. But he confessed that it was not his forte. 'When I was much younger, I tried this equation guessing, and I have seen many students try this, but it is very easy to go off in wildly incorrect and impossible directions.'

As Dirac himself said, 'I think it's a peculiarity of myself that I like to play about with equations, just looking for beautiful mathematical relations which maybe don't have any physical meaning at all. Sometimes they do.'[24] Dirac characterised his technique as 'simply a search for pretty mathematics. It may turn out later that the work does have an application. Then one has had good luck.'[25]

In the search for 'pretty mathematics', Dirac was like an artist, a poet or a novelist, tapping into his unconscious. 'If you are receptive and humble, mathematics will lead you by the hand,' he said. 'Again and again, when I have been at a loss how to proceed, I have just had to wait until I have felt the mathematics lead me by the hand. It has led me along an unexpected path, a path where new vistas open up, a path leading to new territory, where one can set up a base of operations, from which one can survey the surroundings and plan future progress.'[26]

Dirac was struck by the fact that mathematics so perfectly describes nature. 'It seems to be one of the fundamental features of nature that fundamental physical laws are described in terms of a mathematical theory of great beauty and power, needing quite a high standard of mathematics for one to understand it,' he said. 'You may wonder: Why is nature constructed along these lines? One can only answer that our present knowledge seems to show that nature is so constructed. We simply have to accept it.' Dirac went on to speculate: 'One could perhaps describe the situation by saying that God is a mathematician of a very high order, and He used very advanced mathematics in constructing the universe.'[27]

Dirac never quite gave up on the hole theory, and continued to believe in it until at least the 1970s. It did not bother him in the least that it had been comprehensively trashed by his peers. 'That's not a theory,' scoffed Bohr. The truth was that, although the hole theory made little sense to most people, it gave the same results as a modern theory of the electron and so was a great insight. In the words of Dutch Nobel Prize winner Gerard 't Hooft, it was 'Genius!'[28]

As it happens, the hole theory is unnecessary and cannot explain the existence of the antiparticles of subatomic particles known as 'bosons', which unlike electrons are able to crowd into any energy state in unlimited numbers. Antimatter, it turns out, is a generic consequence of combining quantum theory and relativity.

In the modern picture of antimatter, the 'fields' which permeate all of space are the primary things. The most familiar one is the electromagnetic field, which can create or destroy indivisible chunks, or 'quanta', of the field, better known as 'photons' of light (think of light being created by a torch or destroyed by being absorbed by a black cat). Another field is the 'electron field', which, just as the electromagnetic field can create and destroy quanta of the electromagnetic field, can create and destroy quanta of the electron field: electrons and positrons.

Positrons are not as rare as you may think and are naturally emitted by unstable atomic nuclei. Whereas neutron-rich nuclei can achieve stability by turning a neutron into a proton with the emission of an electron, proton-rich nuclei can achieve the same end by turning a proton into a neutron with the emission of a positron. The ejected positrons do not get far before they meet an electron and are annihilated in a puff of high-energy photons, which is why nobody had spotted them before 1932.

Positron-emitting nuclei, however, have proved enormously important in medical imaging. In positron-emission tomography, or PET scanning, a substance containing positron-emitting nuclei is injected into the body. When positrons meet electrons, they create pairs of oppositely directed photons, which can be detected. Since they point back to the location of each annihilation, they can be used by a computer to create a three-dimensional image of the body.

The discovery of the antiproton had to await the advent of a particle accelerator with sufficient energy to create a particle about two thousand times heavier than a positron. It was finally achieved by the 'Bevatron' proton accelerator at the University of California at Berkeley in 1955. A year later came the discovery of the antineutron. And since then, antiparticles of essentially all nature's fundamental subatomic particles have been discovered.

Creating an antiatom, which consists of a positron orbiting an antiproton rather than an electron circling a proton, is a formidable experimental challenge because both antiparticles, once created, must be slowed down hugely before they can combine. But in 1995, physicists at CERN, the European laboratory for particle physics near Geneva, used the Low Energy Antiproton Ring (LEAR) to slow down rather than accelerate antiprotons. By so doing, they managed to bring positrons and antiprotons together, creating nine antiatoms of hydrogen, which each survived for just forty nanoseconds.

Antimatter has the potential to make the perfect rocket fuel because, when antimatter encounters matter, 100 per cent of its mass-energy is converted into other forms. Pound for pound, it therefore packs the biggest punch of any fuel – one hundred times more than a nuclear fuel of equivalent mass. An antimatter rocket therefore need carry only a minimal quantity of fuel; fuel mass is a serious problem for a rocket since it must be boosted along with the rocket itself.

Despite the fact that antimatter powered the Starship Enterprise on its five-year mission to boldly go where no man has gone before, the creation of a real-life antimatter-fuelled spacecraft is fraught with problems. First, the antimatter needs to be stored in such a way that it does not touch the matter of a rocket, which would risk a catastrophic explosion. This might conceivably be achieved by confining the antimatter in a 'magnetic bottle'. Secondly, matter–antimatter annihilation always results in the creation of high-energy photons which, rather than flying out the back of the rocket, which is required to push it forwards, would fly away in all directions.

But the biggest problem in creating an antimatter rocket is accumulating enough antimatter in the first place. So far, we have managed to create only a minuscule quantity of antimatter, and this has taken an enormous effort. If it were possible to make enough antimatter to drive a space probe to Alpha Centauri, our nearest star, a lot more energy would be required to create it in the first place than would be released in its annihilation with matter.

Whether or not antimatter could ever be used for powering an interstellar spacecraft is a minor question. The question of why we live in a matter universe is a deep mystery because all known processes of particle creation, such as pair production, produce equal amounts of matter and antimatter. Dirac said in his Nobel Lecture in Stockholm on 12 December 1933, 'If we accept the view of complete symmetry between positive and negative electric charge so far as concerns the fundamental laws of Nature, we must regard it rather as an accident that the Earth (and presumably the whole solar system), contains a preponderance of negative electrons and positive protons. It is quite possible that for some of the stars it is the other way about, these stars being built up mainly of positrons and negative protons. In fact, there may be half the stars of each kind. The two kinds of stars would both show exactly the same spectra, and there would be no way of distinguishing them by present astronomical methods.'

As Dirac pointed out, antimatter stars would radiate photons just like stars made of normal matter, but he was wrong to say that if our universe contained domains of antimatter intermixed with matter, it would be impossible to tell. Wherever a region of antimatter came up against one of matter, there would be copious annihilation, and astronomers have observed none of the gamma rays expected from this process.

By rights, there should be no universe of either matter or antimatter, only empty space filled with their annihilation products: photons. A clue to why we find ourselves in a universe made entirely of matter comes from the fact that there are about ten billion photons for every particle of matter in the universe. The implication is that, in the Big Bang, there were ten billion and one particles of matter for every ten billion particles of antimatter. After an orgy of annihilation, all antimatter particles were destroyed, leaving one matter particle for every ten billion photons. The key question is: What was the origin of this matter–antimatter asymmetry? Either the fundamental laws of physics are skewed to favour the creation of matter over antimatter or the destruction of antimatter over matter. Exactly how and why they are skewed remains one of the biggest mysteries in modern cosmology.

4

Goldilocks universe

The nitrogen in our DNA, the calcium in our teeth,
the iron in our blood, the carbon in our apple pies
were made in the interiors of collapsing stars.
We are made of starstuff.

CARL SAGAN

As we look out into the Universe and identify the
many accidents of physics and astronomy that have
worked together to our benefit, it almost seems as if
the Universe must have known that we were coming.

FREEMAN DYSON

Kellogg Radiation Laboratory, Pasadena, California, February 1953

The man sitting across the desk was talking utter garbage. Willy
Fowler knew this because he was an experimental nuclear physicist, and nobody in the world could do what this guy was
claiming he could do: predict the precise energy state of a complex atomic nucleus. It was a 'many body' system, in which
numerous protons and neutrons buzzed about each other like a
swarm of submicroscopic bees. Theorists' capabilities were limited to predicting the exact behaviour of a 'two-body' system,
such as an electron circling a proton in a hydrogen atom or the
Moon travelling in its orbit around the Earth.

Nevertheless, here in Fowler's office in Caltech's Kellogg
Radiation Laboratory, a bespectacled Limey astronomer was

claiming that he could do what no nuclear physicist in the world could do. And what was even more outrageous was that his prediction was based not on any consideration of nuclear physics but on an argument the likes of which Fowler had never before heard. 'The universe contains carbon,' he was sure he had heard Fred Hoyle say, 'and therefore a carbon nucleus must have an energy state of exactly 7.65 megaelectronvolts.'*

Hoyle told Fowler he was convinced that the cores, or 'nuclei', of all atoms had been assembled from nuclei of the simplest atom, hydrogen, inside stars that had lived and died before the Sun and the Earth were born. It was, by necessity, a multi-stage process. The first step involved four hydrogen nuclei somehow coming together and forming a nucleus of the second-lightest atom, helium.† The second step was for two helium nuclei to stick together to make a nucleus of beryllium. The problem was that beryllium was unstable and, within a billion-billionth of a second of forming, fell apart. The route to building up heavier

* An electronvolt (eV) is a convenient unit of energy used by physicists. It is the energy acquired by an electron when it is accelerated by a voltage difference of one volt. A megaelectronvolt (MeV) is the energy acquired by an electron accelerated through a voltage difference of a million volts.

† Atomic nuclei contain positively charged particles known as protons and chargeless particles known as neutrons. The two particles, which have essentially the same mass, are together known as 'nucleons'. Hydrogen-1 has one nucleon in its nucleus; helium-4 has four; lithium-6 has six; and so on. Since the protons are balanced by an identical number of electrons orbiting the nucleus, and the electrons determine how an atom connects with other atoms – in short, its fundamental character – the number of protons in a nucleus determines the particular type of atom. Hydrogen atoms contain one proton in their nucleus; helium atoms two; lithium atoms three; and so on. All nuclei except those of hydrogen also contain neutrons, which do not affect the behaviour of an atom but contribute to its mass.

atomic nuclei such as oxygen, calcium and sodium appeared well and truly blocked.

Hoyle claimed there was a way to leapfrog the troublesome beryllium barrier. His scheme, as far as Fowler could tell, required the existence of a high-energy 'excited' state of a carbon nucleus at precisely 7.65 megaelectronvolts above its normal 'ground' state.

Fowler would later recount that his first impression of Hoyle was of someone who had gone a 'long way off his mental compass bearings'.[1] However, working in the shadow of the Mount Wilson 100-inch telescope with which Edwin Hubble had discovered that the universe was expanding in 1929 had made him a nuclear physicist who was tolerant of astronomical ideas. Not showing Hoyle the door would prove to be the smartest career move he ever made.

It was highly probable that Hoyle was wrong, but Fowler adhered to the experimenter's maxim: never close your mind to the unexpected. He called the members of his small research group into his office and made the British astronomer repeat his argument. 'Is there any chance', asked Hoyle, 'that experiments could have missed a 7.65 MeV state of carbon?'

Much of the technical discussion that ensued went well over Hoyle's head, but eventually there was a consensus among Fowler's group. If the state had some very special properties, it was just about conceivable that it could have been missed. Hoyle looked around the assembled faces hopefully, but Fowler shook his head. He had too much work on to carry out an experiment to test Hoyle's outlandish claim. 'Anyone else?' asked Fowler. There was one person. Ward Whaling was a Texan who had recently arrived at Caltech from Rice University in Houston. Turning to Hoyle, he said, 'I'll do it. I'll look for your energy state.'

o   o   o

Hoyle's prediction had had a long period of gestation. It had all begun in the autumn of 1944, when he was a theorist working in England on the development of radar for the war effort. He had been delegated to attend a conference at the end of November in Washington DC. Getting there involved a perilous crossing of the Atlantic, zigzagging to avoid the deadly U-boats. In fact, he had been so worried that, prior to boarding the RMS *Aquitania* at Greenock in Scotland, Hoyle had taken out life insurance at Lloyd's – he had a wife, Barbara, and two young children – and visited his parents in Yorkshire, in case it was the last time he ever saw them. But after ten tedious days at sea along with ten thousand American troops who were returning home, he arrived in the New World.

The bright lights and abundance of New York City were overwhelming after five years of blackouts and rationing in England. Astonished, Hoyle wandered the streets in what seemed to him a 'fairyland', before taking the train south from Pennsylvania Station. In Washington DC, he checked in with the British Embassy and picked up a generous allowance. There were three days to fill before the start of the conference, so he decided to head north to Princeton and see astronomer Henry Norris Russell, famous for his groundbreaking classification of stars in the Hertzsprung–Russell diagram.

Hoyle's interest in astronomy had come about by accident. At Cambridge in 1938–9, he had been the student of Paul Dirac. The story was that the quantum theorist had not wanted a student and Hoyle had not wanted a supervisor, but the pair had been thrust together as a joke by a mischievous faculty member. The famously taciturn Dirac had nevertheless given Hoyle one piece of useful advice. In his view, all the low-hanging fruit of fundamental physics had been picked during the quantum revolution of the 1920s and 1930s. If Hoyle wanted to do important work, he should therefore look for interesting problems in another scientific field.

Hoyle decided to pursue either astronomy or biology, and was luckily saved the trouble of having to choose. Tasked with the responsibility of inviting speakers for a student society, he approached the Cambridge astronomer Ray Lyttleton, who enthused about a particular stellar problem he was working on. Hoyle's interest was immediately piqued. He began collaborating with Lyttleton, and became an astronomer by default.

Hoyle's meeting with Russell at Princeton went well, but it proved more important in terms of what it led to. This would involve assembling a jigsaw puzzle of information gathered from many different sources, illustrating the often chaotic way in which science is done.

On learning that Hoyle would be heading to California after the conference in Washington to visit the US naval headquarters in San Diego, Russell urged him to visit the Mount Wilson Observatory, just north of Los Angeles. He even wrote a letter of introduction to Walter Adams, the observatory's director.

When he got to California, Hoyle went to meet Adams, who immediately sent him up the mountain to spend the weekend at the giant 100-inch Hooker Telescope, the biggest in the world. It was a wonderful opportunity to see the astronomers at work, but it was what happened when the weekend was over that proved crucial. A keen hiker, Hoyle decided to walk down the mountain and was met by Walter Baade in Altadena, the city in the San Gabriel foothills just above Pasadena. The

German–American astronomer was classified as an 'enemy alien' and had been barred from military service, which left him in the enviable position of having unlimited access to the world's biggest telescope while the lights of Los Angeles below were under wartime blackout.

Baade, a scarily clumsy driver despite being a first-class telescope observer, drove Hoyle to his office at Santa Barbara Street. The pair spent a stimulating afternoon discussing the latest developments in astronomy, which culminated in Hoyle leaving with copies of some papers on 'supernovae', prodigiously violent exploding stars that had been discovered by Baade and his Swiss–American colleague Fritz Zwicky. Had Hoyle read the papers immediately they might have meant little to him, but by a turn of fate he did not look at them until he was back in England, by which time he had learnt something that would not only yield a key insight about supernovae but would change the course of his scientific life.

Hoyle had to head to Montréal to hitch a ride on a giant Liberator or Flying Fortress bomber that could fly him non-stop across the Atlantic to Prestwick, near Glasgow. However, bad weather delayed his departure for several days, and while he was waiting he bumped into two physicists he knew from back home. It was an open secret in Cambridge that Nick Kemmer, who had been a student of Wolfgang Pauli, and Maurice Pryce had been recruited by Tube Alloys, the British project to build an atomic bomb.

The idea was to exploit 'nuclear fission', which had been discovered in Berlin by Otto Frisch, Lise Meitner and Fritz Strassmann on the eve of the Second World War. An unstable heavy nucleus was prone to splitting into two, or 'fissioning', and in the process spitting out several energetic neutrons. These could trigger the splitting of further nuclei, raising the possibility of a 'nuclear chain reaction' that would unleash a vast amount of nuclear energy explosively.

Hoyle knew that two different nuclei were capable of fission – a rare 'isotope' of uranium known as uranium-235, and a man-made nucleus, first created in 1940, known as plutonium-239. Making plutonium in sufficient quantities for a bomb would require building a nuclear reactor, or 'pile'. Britain, under bombardment by the Luftwaffe, lacked the resources to follow both routes to a bomb and so had plumped for concentrating uranium-235, a painstakingly slow process that was being carried out at Chalk River, near Montréal. Hoyle took the fact that Kemmer and Pryce were in Canada to mean that sufficient uranium-235 had been accumulated.

As he waited for the weather over Montréal to improve, Hoyle began to wonder about a rumour he had heard that a team consisting of some of the best physicists from America and Europe had been assembled at a secret location in the southwest of the US. It puzzled him; he had thought that it would be easy to create an explosion with uranium-235 by simply slamming together two pieces that in combination exceeded the 'critical mass' necessary to trigger a runaway nuclear chain reaction. The existence of the large team could mean only that for plutonium things were not as simple as that, which would explain why Britain had chosen what he had thought was the more difficult route to develop a bomb.

Clearly, something must prevent two subcritical masses of plutonium from merging, and the only thing Hoyle could think of was the fission of plutonium itself. As two lumps approached each other, he reasoned, fission must generate heat so rapidly that it pushed the pieces apart before a runaway chain reaction could catch hold. If he was right, it would mean that the scientists would have to find a way to force together the pieces of plutonium. As he mused on how they might do that, he realised that the best way would be to cause a spherical shell of plutonium to implode by surrounding it with conventional explosives. Imagining the scenario, he immediately saw a problem: the required implosion would occur only if the shockwave from the explosives was perfectly spherically symmetric, but such a shockwave would be incredibly difficult to engineer. Now he understood why it had been necessary to assemble the high-powered team.

Hoyle's musings were a distant memory when, back in England over Christmas, he finally had time to read Walter Baade's papers on supernovae.[2] The energy released in such a stellar cataclysm was staggering; typically, a supernova outshone an entire galaxy of several hundred billion stars. As he wondered about the energy source, Hoyle realised that only one thing was capable of powering such a detonation: gravity.

If a slate tile falls off the roof of a house, the gravity of the Earth accelerates it so that it hits the ground at high speed. Physicists say that its 'gravitational potential energy' – that is, the energy it possesses by virtue of its location in a gravitational field – is converted into another form: energy of motion. Similarly, if the core of a star shrinks, it is as if the gravity of the star accelerates countless quadrillion slates and their gravitational potential energy is converted into other forms of energy, such as heat. Paradoxically, in a supernova, it is the implosion of the core of a star that drives the explosion of its outer regions into space.

At this point, Hoyle began to put together the jigsaw pieces he had acquired in the US. Just as the implosion of plutonium in a bomb would trigger nuclear reactions, so too would the implosion of the core of a star. The nuclear reactions in each case were entirely different, but that did not matter; the idea that implosion would lead to nuclear reactions was like a light-bulb going on in Hoyle's head.[3] In the inferno of a supernova explosion, those nuclear reactions could potentially forge nature's chemical elements.

The catastrophic shrinkage of the core of the star would be triggered when the core exhausted its fuel and was no longer able to generate the heat needed to stop it being crushed by gravity. Hoyle imagined a frenzy of element-building nuclear reactions going on in the outer envelope of a dying star, driven by the tremendous heat liberated by the shrinkage. Thrust into space by the explosion, the elements would enrich clouds of interstellar gas and dust and, when those clouds fragmented under gravity, would become incorporated into new genera-tions of stars and planets. If Hoyle was right, supernovae were the furnaces in which the elements that make up our bodies were forged.

There are ninety-two naturally occurring elements, ranging from hydrogen, the lightest, all the way to uranium, the heav-iest. It had once been thought that they had all been put in the universe on day one by a Creator, but in the first half of the twentieth century, the idea had arisen that they had actu-ally been made. Scientists had noticed that the abundance or scarcity of each element was related to the nuclear properties of its atoms. For instance, an element whose nuclei were more tightly bound than nuclei of slightly lighter or slightly heavier elements was also more abundant than them, which was a strong hint that nuclear processes had played a key role in the creation of the elements.

The obvious possibility was that the universe had started out with nuclei of the lightest element, hydrogen, and that nuclei of all the heavier elements had been assembled inside stars subsequently, by the repeated sticking together of this basic nuclear building block. In fact, a key discovery Baade made under the blacked-out skies above Los Angeles was that the Milky Way contains two distinct populations of stars. In the 'spiral arms', where the Sun orbits, are hot blue stars with a relatively high concentration of heavy elements, and in the centre of the galaxy are cool red stars with a low concentration.* As would later be shown, the blue 'Population I' stars are young and the red 'Population II' stars are old, their heavy-element concentrations revealing that heavier elements have become more common as the galaxy has aged, exactly as would be expected if heavy elements are built up inside stars over time.[4]

The building of ever-bigger nuclei is not easy because it requires forcing together ever more protons, and like charges repel each other ferociously. The only way the repulsion can be overcome is by the slamming together of nuclei at ever-greater speeds, which, since temperature is a measure of microscopic motion, is synonymous with ever-greater temperature. In fact, the building of heavy elements requires the existence of a furnace at a temperature of many billions of degrees.

It was the belief that the interiors of stars could never attain such mind-bogglingly high temperatures that caused the

---

* Atoms of a particular element absorb and emit light at only certain wavelengths, which act as a fingerprint for the element. It is by means of such fingerprints that different elements reveal themselves to astronomers in the light of stars. The wavelengths correspond to the energy absorbed or shed by electrons moving between different orbits inside an atom.

American physicist George Gamow to search for an alternative furnace for forging the elements and claim that a hot Big Bang fitted the bill. But in 1944, reading Baade's papers, Hoyle saw an opportunity to demonstrate that there was no need to look for an alternative furnace; if he was right, the interiors of stars could reach temperatures at least a thousand times as great as the ten million degrees or so at the heart of the Sun.

The sequence of nuclear reactions that built up the elements inside stars was likely to be complex, and Hoyle did not have the slightest idea of its details. However, the beauty of a super-nova, he realised, was that the inferno was so preposterously dense and hot that the details did not matter. In the ensuing submicroscopic frenzy, nuclei would constantly form and break apart, and a balance would be reached in which the processes of creation and destruction perfectly matched. Such a balance depended only on how tightly bound each nucleus was, and in such a state of 'statistical thermodynamic equilibrium', the relative abundances of the elements would become fixed and unchanging. In the jargon, they would 'freeze out'.

All Hoyle needed to know was the abundances of different elements and how tightly bound were their nuclei. Unfortunately, his radar work had left him in the West Sussex countryside, with no access to data of this kind. Then, in March 1945, his research took him to Cambridge, where he ran into Otto Frisch. The Austrian physicist had recently returned from the US, where he had been working with the team developing the atomic bomb at Los Alamos in New Mexico. Frisch, it turned out, had exactly what Hoyle wanted: from the drawer in his desk, he pulled out a table of nuclear data, painstakingly compiled by the German nuclear physicist Josef Mattauch.

From Cambridge University Library, Hoyle borrowed a book written by Victor Goldschmidt. In 1937, the Swiss-Norwegian physicist had carried out a pioneering study of the make-up of the universe, pulling together data from the Earth's crust, the Sun and from meteorites. The table in which he summarised his results revealed which elements were common and which were rare.

With Goldschmidt and Mattauch's data, Hoyle had everything he needed. For a range of different temperatures, he calculated the relative abundances of the elements that would freeze out in nuclear thermodynamic equilibrium, and he discovered something striking: at a temperature of between two and five billion degrees, the relative abundances he predicted for copper and nickel, cobalt and chromium – the elements on which our modern civilisation is based – matched exactly those found by Goldschmidt. Hoyle was euphoric. He now had quantitative proof that such 'iron group' elements had been forged in supernovae.[5] 'All humans are brothers,' as the American astronomer Allan Sandage would one day put it. 'We came from the same supernova.'

Hoyle finally had his proof that stars had forged some of nature's elements. Their interiors were, after all, capable of achieving the enormous temperatures and densities that were necessary. But he believed that it was not just *some* elements that had been created inside stars but all of them. He was a long way from proving that, but crucially he now had proof that stars could achieve the necessary extreme conditions for 'nucleosynthesis'.

The reason people had thought stars could not achieve such conditions – prompting Gamow to look to the Big Bang as an alternative crucible for the building of the elements – was because of an uncharacteristic error made by Arthur Eddington. It was the English astronomer's detection of light-bending by the gravity of the Sun in 1919 that had simultaneously shown Newton to be wrong and transformed Einstein into a scientific superstar. By the 1930s, astronomers had guessed that starlight was a by-product of the 'fusion' of the nuclei of hydrogen into the nuclei of helium.* However, Eddington believed that the helium 'ash' would become mixed throughout a star, gradually diluting its hydrogen fuel and extinguishing the nuclear reactions. Evidence that he might be wrong could be seen in the night sky in stars like Betelgeuse, in the constellation of Orion; such 'red giants', far from fading, typically pumped out ten thousand times as much heat as the Sun.

Making sense of such stars was the problem that had piqued Hoyle's interest when he had first met Lyttleton. The pair of them had realised that, if a star becomes non-uniform in composition, rather than staying well mixed, as Eddington believed, it automatically becomes hotter and denser, which might explain the light output of a red giant. Hoyle and Lyttleton imagined a star achieving such a non-uniform state by flying through a cloud of interstellar gas and accumulating an outer mantle of hydrogen.

---

* In fact, Gamow had provided this vital ingredient. The first person to apply quantum theory to the atomic nucleus, he had discovered in 1928 that, in the 'alpha decay' of a heavy element such as radium, an alpha particle, or helium nucleus, can escape the nucleus, even though it appears to have insufficient energy to do so. This phenomenon of 'quantum tunnelling' is possible because the quantum wave associated with the alpha particle extends outside the nucleus, giving the alpha particle a small probability of being found there at any time. In 1929, Robert Atkinson and Fritz Houtermans turned Gamow's idea on its head, showing how inside the Sun one nucleus could tunnel into another, despite their ferocious mutual repulsion appearing to make it impossible. As a by-product, such a nuclear reaction created sunlight.

This turned out to be unnecessary when Eddington discovered his error.[6] The mechanism that he believed mixed helium evenly throughout a star was nowhere near as efficient as expected. Consequently, a star's helium, being heavier than hydrogen, fell to its centre, where it heated up like any gas that is compressed. As they evolved, stars automatically acquired non-uniform interiors, their cores becoming ever denser and hotter.

As a star built up heavier and heavier elements, and each fell to its centre, it would develop an internal structure reminiscent of an onion, with each successive layer denser and hotter than the one surrounding it. This was perfect, Hoyle realised, for forging all the elements. When such a star blew up as a supernova or lost matter in a 'stellar wind', some of those elements would end up in the interstellar medium as raw material for the next generation of stars.

In Gamow's Big Bang furnace, there was only a window of opportunity between about one minute and ten minutes after the birth of the universe in which elements might be forged; after that, cosmic expansion made the fireball too rarefied and too cool. The scheme was able to forge only helium and a few of the very lightest elements. Stellar furnaces, by contrast, had billions of years available for working their alchemical magic. With so much time available, it was patently obvious that stars would win out over the Big Bang. Or would they?

Gamow's scheme failed not only because there was less than ten minutes to build up heavy elements in the furnace of the Big Bang but for an even more fundamental reason: in nature, there is no stable nucleus of mass 5 or 8.

Both protons and neutrons – collectively known as 'nucleons' – existed in the fireball of the Big Bang, though neutrons decayed into protons, or hydrogen nuclei, by the time the universe was a little over ten minutes old. A nucleus of the second lightest element, helium, consists of four nucleons – two protons and two neutrons – and so must be built up in several steps. Once helium-4 formed in the Big Bang, the obvious route to building heavier elements was to add another nucleon to make a nucleus of mass 5, or stick together two helium-4 nuclei to make a nucleus of mass 8. But the absence of stable nuclei of mass 5 and 8 in nature meant that the road was blocked, which was as fundamental a problem for the furnaces of stars as for the furnace of the Big Bang.

After his supernova revelation, Hoyle's work on element synthesis in stars was therefore stymied, so he turned to cosmology, the science of the large-scale universe. In 1948, together with Hermann Bondi and Tommy Gold, he proposed the 'steady-state theory'. Edwin Hubble, observing from the Mount Wilson Observatory, had in 1929 discovered that the universe was expanding, its constituent galaxies flying apart like pieces of cosmic shrapnel. According to the steady-state theory, as the galaxies recede from each other, new material fountains into existence in the gaps and congeals to form new galaxies. Although at first sight the idea seems ridiculous, it is no more ridiculous than the idea of all matter erupting into existence in one go in a Big Bang, and it has the advantage that the universe on the large scale looks the same at all times. Such a universe can have existed forever, since only by changing can a universe have an origin. There is no need answer the question: How did it all begin?

It was partly because of his interest in cosmology that Hoyle attended a meeting of the International Astronomical Union in Rome in the summer of 1952. There, he found himself in the audience of a session on 'extragalactic nebulae', or galaxies, being chaired by Walter Baade. The Caltech astronomer had carelessly overlooked the need for a secretary to take the Commission's minutes, so he asked Hoyle to help out. During the session, Baade presented sensational evidence that the universe was twice as old as had been estimated by Hubble. When, months later, an astronomer who had been in the audience that day stole Baade's conclusion and passed it off as his own, Hoyle saved the day: his minutes proved that Baade had been outrageously ripped off and ensured that he received the credit he deserved.

Baade sat on the combined astronomical steering committee of the Mount Wilson Observatory and the California Institute of Technology, which almost certainly explained why, in the autumn of 1952, Hoyle received an invitation to spend three months at Caltech. He jumped at the chance and arrived in Pasadena thinking about nucleosynthesis in stars and possible ways of leapfrogging the troublesome mass 5 and mass 8 gap. Caltech was the perfect place to be; it had both a world-class astronomy department and an active nuclear physics group.

Research into nuclear physics had begun at Caltech shortly after the construction of the Kellogg Radiation Laboratory in 1930–1. Built with money from Will Keith Kellogg, 'The Cornflake King', the laboratory was initially equipped with a powerful 1 MeV X-ray tube, to study not only the physics of such radiation but its application in the treatment of cancer.[7] But when John Cockcroft and Ernest Walton sensationally split the atom with high-speed protons in Cambridge, England, in 1932, Charles Lauritsen, the lab's director, immediately changed the direction of its research.

An X-ray tube uses a high voltage difference to accelerate electrons so they smash into a metal target, creating high-energy

X-rays in the process. It was straightforward to adapt Kellogg's X-ray tube and use its high voltage difference to instead accelerate particles such as protons and smash them into atomic nuclei. By observing the resultant shrapnel, the physicists at Kellogg were able to measure the speed of nuclear reactions that transformed one kind of nucleus into another. In fact, when Hans Bethe had proposed the so-called 'CNO cycle' of nuclear reactions for turning hydrogen into helium inside stars and generating starlight as a by-product, it was Willy Fowler and his team at Kellogg that had measured the speed of the cycle's individual nuclear reactions. They discovered that it operated efficiently only at temperatures much higher than the central temperature of the Sun, ruling it out as the principal power source of all but the most massive stars.*

Fowler had considered himself merely a nuclear physicist, and it was a revelation when Bethe taught him that what he was doing in the lab might be mimicking the energy-generating nuclear reactions deep inside of stars. Still, it took a young theorist from Cornell University in 1951 to make Fowler realise that his team might also be able to mimic element-building nuclear reactions in stars. Ed Salpeter raised the possibility that the mass 5 and mass 8 barrier might be bypassed by an exceedingly unlikely nuclear process.

What if three helium nuclei – commonly known as 'alpha particles' – came together simultaneously inside a red giant star to create a nucleus of carbon-12? Imagine three people in a supermarket car park crashing their shopping trolleys into each other simultaneously. You would have to wait a long time to see

---

* As would later be discovered, lower-mass stars like the Sun are powered by another sequence of nuclear reactions, known as the 'proton–proton chain'.

such an event, but one thing stars have in abundance, realised Salpeter, is time – millions or even billions of years, compared with the ten minutes or so in which element-building must be completed in the Big Bang.

Not surprisingly, Salpeter's 'triple-alpha process' did not work – it was so infrequent that it produced only the tiniest of quantities of carbon. Yet it was an observational fact that carbon is very abundant cosmically – it is the fourth most common element in the universe after hydrogen, helium and oxygen.

When Hoyle arrived at Caltech at the end of 1952, he was aware of Salpeter's work, and was sure the Cornell theorist was correct that the only conceivable way to leapfrog the mass 5 and mass 8 barrier was for three helium nuclei to collide and stick together. The question was therefore: Was there any way to speed up Salpeter's process? Hoyle was certain there was, and he had a wild idea about how to do it.

Predicting the manner in which the nucleons buzzed about inside a nucleus was beyond the capabilities of any theorist. However, some internal configurations of nucleons were more stable than others, because it was an observable fact that each nucleus could exist in one of a number of 'energy states'. It had, for instance, a lowest possible energy, or 'ground', state, and a number of higher energy, or 'excited', states, arranged above it like the rungs of a ladder.

If there existed an excited state of carbon-12 at precisely the energy of three helium nuclei at the 100-million-degree temperature at the heart of a red giant, Hoyle thought, it would cause the nuclear reaction between three helium nuclei to be 'resonant'. In exactly the same way that a child's swing pushed at its natural, or resonant, frequency speeds up, the nuclear reaction would be boosted. Hoyle carried out the relevant calculations and found that, if the triple-alpha reaction to make carbon-12 was resonant, it would be faster than that calculated by Salpeter, not by a factor of ten or one hundred or even one thousand but by an astonishing factor of ten million. Most importantly, Hoyle's back-of-the-envelope calculations showed that such an enhancement was capable of explaining the abundance of carbon in the universe.

The energy of three helium nuclei at the 100-million-degree temperature inside a red giant was about 7.65 MeV. For the nuclear reaction to make carbon-12 to be resonant, carbon-12 must therefore have an energy state at precisely 7.65 MeV above its ground state. But did it? This was the question that Hoyle asked Fowler in his office that day in February 1953 – the question which had, fortunately for Hoyle, piqued the interest of Ward Whaling.

Kellogg Radiation Laboratory, Pasadena, California, February 1953

Whaling was already an admirer of Hoyle's chutzpah. On 30 December 1952, shortly after his arrival in Pasadena, the British astronomer had given a public talk on his steady-state theory at the midwinter meeting of the American Physical Society at Caltech. It had created such excitement in the Los Angeles area that it had to be moved to a bigger auditorium at Pasadena Junior College. The talk had so impressed Whaling that he had started attending lectures that Hoyle gave each week in the Robinson Laboratory of Astrophysics, just a few minutes' walk from Kellogg.

In his lectures, Hoyle developed his ideas on how element-building might proceed inside stars. Hoyle would invariably be shot down with a killer objection from one of the astronomy department's big beasts, but it seemed as if criticism from the likes of Jesse Greenstein and Fritz Zwicky was like water off a duck's back to Hoyle. The following week he would be back with an inventive way of circumventing their objection, only to be shot down again. Whaling found it exhilarating; Hoyle's knowledge of astronomy and nuclear physics was ropey, but this was compensated for by his exceptional mathematical ability and powerful imagination. Above all, he was eager to learn, and by bouncing back and forth between the nuclear physicists in Kellogg and the astrophysicists in Robinson he learnt fast.

It was Whaling's admiration for the bespectacled Yorkshireman that caused him to pipe up and volunteer to look for the excited state of carbon-12. That and the fact that, unlike Fowler, he was not snowed under with other work.

Whaling's plan was to use the Kellogg accelerator to fire 'deuterons' at nuclei of nitrogen-14. A deuteron is a nucleus of deuterium, or 'heavy hydrogen', and contains a proton and a neutron, while a nucleus of nitrogen-14 contains seven protons and seven neutrons. Each collision would result in the creation of carbon-12 and helium-4 nuclei. The key would be to measure the energy of the helium nuclei: the available energy would be shared between the carbon-12 and the helium-4, which would mean that, if the carbon-12 was created in its low-energy, ground state, the helium would be left with a relatively large amount of energy. However, if the carbon-12 was created in a high-energy, excited state, it would leave the helium nuclei with relatively little. Evidence for Hoyle's predicted state would be the detection of some helium nuclei with precisely 7.65 MeV less energy than the rest.

As explained in the previous chapter, the energy of a nucleus can be measured by observing how much its trajectory is bent by a strong magnetic field; those nuclei with the highest energy are bent the least and those with the lowest energy the most. A suitably strong magnet was available; the problem was that it was not in the same room as the particle accelerator, and it weighed several tonnes.

Whaling's team consisted of his graduate student, Ralph Pixley, a postdoc called Bill Wenzel and a visiting Australian postdoc called Noel Dunbar. None of them could think how to transport the magnet the thirty metres or so down the narrow corridor. Fortunately, their resident engineer, Vic Ehrgott, came up with the ingenious idea of manoeuvring the magnet onto a steel plate resting on hundreds of tennis balls.[8] With the weight spread over such a large number of balls, no single ball was crushed flat.

One member of the team had the job of retrieving tennis balls spat out at the rear of the metal plate and tossing them to the front, where someone else stuffed them under the plate. With a great deal of effort from several red-in-the-face and groaning people, it was possible to inch the load forward. A similar technique using wooden rollers had been used thousands of years before by the workforce of the Egyptian pharaoh to transport blocks of stone from a quarry to the site of the pyramids, and its modern-day incarnation proved equally effective. After two days, Whaling and his colleagues finally had the magnet in the same room as the accelerator and the experiment was ready to begin.

o   o   o

For ten days, Hoyle was on tenterhooks.[9] Each day, he left his office in Robinson and walked the short distance through the winter sunshine to the Kellogg Lab. To his left he could see the tiny dome of Mount Wilson Observatory high in the San Gabriel Mountains and smell the faint tang of oranges in the air. It was quite a contrast when he plunged into the gloom of the lab and saw Whaling and his team beavering away, surrounded by a jungle of power cables, transformers, whirring vacuum pumps and diving bell-like chambers, in which atomic nuclei were fired at each other.

Having his prediction tested felt to Hoyle like being in the dock with his life in the balance, while the jury was out deliberating. The difference was that a prisoner knows whether they are innocent or guilty; if they are innocent, they hope the jury gets it right, and if they are guilty, they hope they get it wrong. The jury of experimentalists, however, is always right. 'The problem is you don't know whether you're innocent or guilty, which is what you stand there waiting to hear, as the foreman of the jury gets up to speak,' said Hoyle.

On the tenth day, Whaling was waiting for Hoyle. He pumped his hand and gushed his congratulations; the prediction had been borne out. Unbelievably, there was an energy state of the carbon-12 nucleus at 7.68 MeV, which was compatible with 7.65 MeV within the range of experimental error. With the way to bypass the mass 5 and 8 barrier now established, the route to building all heavier elements lay open. Hoyle's outrageous prediction had been proved right; he had peered into the heart of nature and spied something that mere mortals – or, at least, theoretical nuclear physicists – had been unable to see. 'The day I heard the result, the scent of orange trees smelled even sweeter,' he said.[10]

'It was really quite a tour de force,' said Fowler. 'A man walked into our lab and predicted the existence of an excited state of a nucleus, and when the appropriate experiment was performed it was found. No nuclear theorist starting from basic nuclear theory could do that. Hoyle's prediction was a very striking one.'[11]

But what compounded Fowler's amazement was the manner of Hoyle's prediction. He had predicted the 7.65 MeV energy state of the carbon-12 nucleus using an unprecedented argument: it had to exist because if it did not, the universe would contain no carbon or heavy elements. Nobody in the history of physics had ever used so preposterous an argument to make such a precise prediction about the world. Among the brotherhood of the magicians, Hoyle has a unique place.

o   o   o

When Hoyle had time to ponder the discovery of the 7.65 MeV energy state of the carbon-12 nucleus, he began to appreciate how the existence of the heavy elements out of which we are made appears to be dependent on not one but several pieces of remarkably good fortune. The first is the non-existence of a stable state of a nucleus of beryllium-8. The second is the existence of an excited state of a carbon-12 nucleus at precisely 7.65 MeV. But there is also a third piece of nuclear luck.

There is no energy state of an oxygen-16 nucleus at the combined energy of a carbon-12 and a helium-4 nucleus at the 100-million-degree temperature inside a red giant; if there was, the conversion of carbon-12 into oxygen-16 would be resonant. In other words, the instant carbon-12 was made in the triple-alpha process, it would immediately be converted into oxygen-16. The universe would end up with no carbon what-soever, whereas in reality it contains roughly equal amounts of carbon and oxygen.

In 1973, the Australian physicist Brandon Carter popularised the idea that many of the 'fundamental constants' of nature, such as the strength of the electromagnetic force and the mass of the electron, have the values they have because, if they did not, it would be impossible for stars, planets and life to exist. In other words, the fact that we are here is a key observational fact. After all, if things were not the way they are, we would not be here to remark on the fact.

Not surprisingly, the topsy-turvy logic of this 'anthropic principle' has proved controversial. It does not help the idea's credibility that its proponents have pointed out that the electromagnetic or gravitational forces have the strength they have only *after* observing the consequences of those forces in the universe. Hoyle's prediction of the 7.65 MeV energy state of the carbon-12 nucleus is unique in that it was made in advance of any observation or experiment. And in the years since 1973, it has been hailed as the big success of the anthropic principle.[12]

The three pieces of nuclear good fortune may not actually be as necessary for our existence as they seem at first sight. Proponents of the anthropic principle point out that, if the strong nuclear force which binds the nucleons inside nuclei were a few per cent weaker, it would be impossible to make sufficient carbon-12, but it is not often pointed out that, if the strong nuclear force were a little stronger, it would make the nucleus of beryllium-8 stable. Crucially, this would open up an entirely new route to the building of carbon-12 and all heavier elements. So at best, the fact that beryllium-8 is unstable is a one-sided piece of luck.

In 1953, Hoyle in effect said, 'Heavy elements exist, therefore there must exist a state of carbon-12 at an energy of 7.65 MeV to open the door to the building of heavy elements.' However, he eventually saw things in anthropic terms and his statement metamorphosed into 'I exist, therefore there must exist a state of carbon-12 at an energy of 7.65 MeV.'

The bypassing of the beryllium barrier opened the road for the building of heavy elements. As a massive star evolved and its core became ever denser and hotter, helium nuclei deep in its interior would stick to oxygen-16 to make neon-20; helium nuclei would stick to neon-20 to make magnesium-24; and so on. This 'alpha process' would culminate in the addition of helium to silicon, to make iron at a temperature of about three billion degrees. At this juncture, things would go awry for the star; 'silicon burning', unlike the preceding nuclear reactions, does not liberate energy but sucks it out of a star. And, since the heat from such nuclear energy is what provides the outward push that prevents gravity crushing a star, the core implodes. This is the process – which is still not well understood – that results in the ejection of the outer envelope of a star as a 'supernova', spraying into space many of the elements it has painstakingly built up over its lifetime.

The iron-group elements are created in the nuclear thermodynamic equilibrium that exists briefly in the supernova explosion, but there are many other processes such as the alpha process that are responsible for building up elements. In fact, in a monumental paper published in 1957 by Margaret and Geoffrey Burbidge, Fowler and Hoyle, and universally known as 'B²FH', eight distinct element-building processes were identified as producing the elements we see in the universe today.[13] Two processes – the rapid and the slow-neutron processes

– make nuclei by adding neutrons one at a time. Since neutrons have no electric charge, such processes overcome the problem of charged nuclei repelling each other before they can get close enough to stick. The rapid and slow-neutron processes make neutron-rich nuclei in supernovae explosions and red giant stars, respectively.

Willy Fowler won the 1983 Nobel Prize in Physics for figuring out why elements such as iron and nickel are common and elements such as lithium and beryllium are rare. Hoyle did not share the prize, although, as Fowler later remarked, he himself would probably have remained a run-of-the-mill nuclear physicist had it not been for the visit of Fred Hoyle to his office on that fateful day in the winter of 1953.

Despite all the successes of B^2FH, the origin of a small number of elements such as gold and silver was mysterious, until recently. The puzzle was finally solved only on 17 August 2017, when gravitational waves were detected by the Laser Interferometric Gravitational-Wave Observatory, or LIGO. They came from the merger of two super-compact 'neutron stars'. The gamma rays picked up on Earth carried the 'fingerprint' of gold and silver, and revealed the creation of a quantity of gold equivalent to twenty times the mass of the Earth.

The extraordinary story begun by Hoyle in 1944 has reached its final chapter. We are more connected to the stars than even the astrologers guessed. Would you like to see a piece of a star? Hold up your hand. The iron in your blood, the calcium in your bones, the oxygen you take in with every breath were all forged inside stars that lived and died before the Earth and Sun were born. You are stardust made flesh. You were literally made in heaven.

5

Ghost busters

Neutrino physics is largely an art of
learning a great deal
by observing nothing.

HAIM HARARI[1]

I have done a terrible thing: I have postulated a
particle that cannot be detected.

WOLFGANG PAULI

Savannah River, South Carolina, 14 June 1956

Frederick Reines was singing as he drove to the bomb plant. He
loved to sing almost as much as he loved to do physics. Back in
college in New Jersey, he had even taken lessons from a voice
coach at the Metropolitan Opera and sung solos in Handel's
Messiah.[2] When he was working on a particularly tough theoretical problem, he had been known to sing for hours on end
while locked away in his office. But on this June morning there
was a very specific reason why his fine baritone boomed out
through the wound-down car window, turning the heads of
pedestrians walking by on the sidewalk. After almost a year of
exhausting work – five years, if you counted the total effort that
had led up to this day – he was in celebration mode. He and his
team were about to achieve the impossible.

It was eight miles from Aiken, the pretty beach-side community where they had been domiciled since November, to the

Savannah River Plant. As he drove out of town, the sweet smell of camellias and magnolias came in through the window on the hot damp air, reminding him of how exotic South Carolina had seemed when they had arrived from the high desert of Los Alamos. On their first drive out from Aiken through the swampy Savannah River Valley, their car lurched over something in the road and they were tossed about like rag dolls. Looking back, they saw that what they had assumed was a speed bump was actually a giant rattlesnake.[3]

At the gates to the Savannah River Plant, Reines pulled up behind a long double-line of cars. The site, with its five nuclear reactors, separation facilities and waste dumps, covered an area larger than New York City and employed almost forty thousand people. When the US government had announced its plan to build the facility, it had said it was not for the 'manufacture of atomic weapons', but that was splitting hairs. Everyone knew the truth: it made the fuel for nuclear weapons – which is why, even in the shops and beach bars of Aiken, Savannah River was referred to as the 'bomb plant'.[4]

In early September 1949, a US Air Force B-29 bomber had sniffed the air high above the Pacific coast of the Soviet Union and caught the unmistakable aroma of an atomic bomb blast. Like most of his colleagues at Los Alamos, Reines had worked on the Manhattan Project to build the first atomic bomb, and he still remembered his shock at the announcement, a mere four years after Hiroshima, that the Russians had caught up and the US no longer had an atomic monopoly.

To counter the Soviet threat, President Harry Truman embarked on a drive to build a 'superbomb' whose destructive power would dwarf an atomic bomb. It involved the construction of vast facilities across the country, not only to make the fuel for such 'hydrogen bombs' but to assemble them. As part of the programme, on 28 November 1950 the US government announced the seizure of almost 500 square kilometres of land by the Savannah River to make two key components of nuclear bombs: tritium and plutonium. Four towns were bulldozed and six thousand people moved, and, by early 1952, the plant was in full production mode.[5]

On 1 November 1952, the US exploded a hydrogen bomb on Elugelab, part of Enewetak Atoll, a Pacific island liberated from the Japanese in the Second World War. With 700 times the destructive power of the bomb dropped on Hiroshima, it vaporised the island, creating a radioactive mushroom cloud 150 kilometres across and gouging a hole in the ocean floor more than two kilometres wide and as deep as a sixteen-storey building. But a mere nine months later, in August 1953, came the scarcely believable news that the Russians had detonated their own hydrogen bomb. Their design could not be scaled up to make bigger explosions, but everyone knew it was only a matter of time. Sure enough, on 22 November 1955, at the Soviet test site at Semipalatinsk in Kazakhstan, the Russians exploded their first true hydrogen bomb.

Reines reached the head of the line of cars, flashed his ID card through the open window and accelerated towards the hulking shape of P Reactor. The Savannah River facility boasted five reactors – R, P, K, L and C – whose letter designations had been chosen entirely at random. They were built at two-and-a-half mile intervals, so they could not be wiped out by a single Soviet nuclear strike, and were spread along a horseshoe-shaped curve to make them immune to a straight-line bombing run. Each reactor rose sixty metres into the air and was sunk twelve metres into the ground, for even more protection. It was this last feature that was of key importance to Reines and his team. It was what had brought them from New Mexico in search of their impossible quarry: a ghostly subatomic particle that had been predicted a quarter of a century earlier and whose existence would almost certainly be confirmed that day.

Zurich, December 1930

The elusive particle had been predicted by the Austrian physicist Wolfgang Pauli. A one-time infant prodigy, Pauli had at twenty-one written such a masterful survey of the theory of relativity that it had astonished even the theory's creator, Albert Einstein. In fact, Pauli – his confidence bordering on arrogance – had famously stood up at the end of a lecture given by Einstein, turned to the audience and reassured them that 'What Professor Einstein said is not entirely stupid.'[6]

In the mid-1920s, Pauli was one of the principal architects of 'quantum theory', the revolutionary description of the sub-microscopic world of the atom and its constituents. His name is immortalised in the 'Pauli exclusion principle', which, by preventing electrons from piling on top of each other, makes atoms and the everyday world possible.

By the late 1920s, a new puzzle began to worry Pauli and his peers. It concerned radioactive 'beta decay'. A beta particle was one of the three distinct types of radiation spat out by the nucleus of an unstable atom as it 'decayed', rearranging its constituents to attain a more stable state. In 1899, three years after the discovery of radioactivity by Frenchman Henri Becquerel, New Zealand physicist Ernest Rutherford had shown that beta particles were 'electrons' – not the common-or-garden variety that orbited the nucleus of an atom, but electrons from inside the nucleus.

In the world of the atomic nucleus, greater stability is synonymous with lower energy. Consequently, when a nucleus decays, it drops from a higher to a lower energy 'state'. The excess energy is spat out as an alpha particle, beta particle or gamma ray. Experimenters observed that alpha particles and gamma rays were emitted at precise energies, which made perfect sense if those energies were equal to the difference in energy between the initial and final states of the nucleus. However, the English physicist James Chadwick discovered something peculiar about beta particles in 1914: unlike their cousins, they were emitted not with a precise energy but with a continuous range of energies.

Think of a gun, which uses a fixed amount of energy to fire a bullet. Every bullet exits the gun at the same speed. It is never the case that one leaves at moderate speed, the next at high speed and the one after that so slowly that it dribbles out the end of the gun barrel. But this is precisely what the tiny electron bullets spat out in beta decay do. Not surprisingly, physicists were shocked at what Chadwick's experiment was telling them.

This behaviour of beta particles might, of course, have a perfectly mundane explanation. Perhaps before escaping they bounced around inside an atom like a ball bearing in a pinball machine, striking multiple electrons and losing a portion of their energy to each one. However, by 1927, this possibility had been ruled out by an experiment by Charles Ellis and William Wooster at Cambridge University.[7] The beta particle puzzle remained, and was so serious that it caused Niels Bohr, one of the founding fathers of quantum theory and the greatest physicist of the twentieth century after Einstein, to question one of the foundation stones of physics – that energy can neither be created nor destroyed but only transformed from one type to another. Perhaps in the world of the atom, Bohr suggested, processes do not obey the 'law of conservation of energy'.

Enter Pauli, a physicist at the Swiss Federal Institute of Technology in Zurich. To him, the conservation of energy was like a life raft in a violent, storm-tossed sea, and abandoning it was absolutely unthinkable. 'Bohr is on entirely the wrong track,' he said. But what, then, was the solution to the beta particle puzzle?

Pauli was having the worst year of his life. Two years earlier, in November 1927, his mother, having been abandoned by her husband, committed suicide. The event had such a profound effect on Pauli that he left the Catholic Church, no doubt feeling abandoned by God. Then, on 23 December 1929, he married Käthe Deppner, a twenty-three-year-old cabaret dancer from Berlin six years his junior. When she met Pauli she was seeing a chemist called Paul Goldfinger, and she continued the affair during their marriage. An anguished Pauli, who was not even living with his wife, told a friend that he was only 'loosely married'.[8]

Losing his wife to another man hurt, but Pauli felt the humiliation even more keenly because it affected his pride. 'Had she taken a bullfighter I would have understood,' he complained to friends. 'With such a man I could not compete – but a chemist – such an average chemist!'[9]

Pauli's troubled marriage to Deppner resulted in him developing a drink problem and a smoking habit.[10] 'With women and me things don't work out at all,' he wrote despairingly. 'This, I am afraid, I have to live with, but it is not always easy.

I am somewhat afraid that, in getting older, I will feel increasingly lonely.'[11, 12]

In the darkest times, occupying his mind with the problems thrown up by quantum theory may have served as an escape from his troubles, but this may have further strained his relationship with Deppner. She reported that Pauli received many letters from physicists, especially quantum pioneer Werner Heisenberg, and would walk around in their apartment 'like a caged lion . . . formulating his answers in the most biting and witty manner'.[13] It was during the eleven anguished months that he was loosely married to Deppner that Pauli came up with the idea for solving the puzzle of beta decay.

Pauli set out his solution to the problem on 4 December 1930, in an open letter to fellow scientists at a meeting in Germany.[14] 'Dear Radioactive Ladies and Gentlemen,' it began. 'Unfortunately, I cannot appear personally in Tübingen, since I am indispensable here in Zurich because of a ball on the night of 6 to 7 December.' The dance was at the 'Baur au Lac', the most distinguished hotel in the centre of Zurich, and it was a mere ten days since his divorce. Emotionally bruised though he was, Pauli intended to get straight back on the horse and find himself another woman.

The letter was read out aloud to attendees at the Tübingen meeting, including Lise Meitner, who would later play a crucial role in the discovery of 'nuclear fission'. Pauli pointed out that even if a fixed amount of energy was available in beta decay, the fact that the electron emitted from the nucleus did not have a fixed amount could be explained if it shared it with a hitherto unknown particle.

Think of the gun again. If a bullet emerged from the barrel with a second projectile, the two would share the available energy. If the second projectile took very little of the energy and the bullet took the lion's share, it would be expelled at high speed. If the second projectile took most of the energy and the bullet had very little energy, it might emerge at such a low speed that it dribbled out the end of the gun. Depending on how much of the available energy was used by the second projectile, the bullet could have any of a range of possible energies.

However, no second particle had been identified accompanying the electron emitted in beta decay. Pauli's new particle must therefore interact very rarely with the atoms of normal matter, and he estimated that it would take a ten-centimetre-thick lead wall to stop it in its tracks.

On the hypothetical particle's other properties, Pauli was also quite specific. In order for it to not noticeably affect the mass of a nucleus, it must weigh very little, if anything at all. He did not realise that it might not actually exist in the nucleus but instead be created at the moment of emission, just as a photon of light is created at the moment of emission and is in no sense taken from a pre-existing 'bag of photons' within an atom. Pauli was also specific about the electric charge of the hypothetical particle, which, like energy, cannot be created or destroyed. In beta decay, for instance, there is no net change in the total charge – though the nucleus increases its positive charge, this is compensated for by the negative charge carried by the emitted electron.* In order for the new particle not to upset this delicate balance, it must therefore carry no charge.

* We now know that in beta decay, a 'neutron' in a nucleus changes into a proton. Since both protons and neutrons are composite particles made of triplets of 'quarks', we can be more specific: a down-quark in a neutron changes into an up-quark, turning the neutron into a proton.

In recognition of its electrical neutrality, Pauli christened it a 'neutron', a name that would later be changed to 'neutrino'.

'I don't feel secure enough to publish anything about this idea,' Pauli wrote in his letter to the Tübingen meeting. The neutrino was a 'desperate remedy'. The reason was that, in 1930, only three subatomic building blocks of matter were known: the 'proton' in the nucleus of the atom; the electron, which orbited the nucleus; and the photon, the particle of light. By adding another particle, Pauli was increasing the number of nature's fundamental building blocks by a third.

The first time Pauli announced the neutrino in public was on 16 June 1931 at the inaugural summer meeting of the American Physical Society in Pasadena, but it gained more traction among physicists four months later, at a meeting in Rome organised by Enrico Fermi. Fermi, who would turn out to be the greatest Italian scientist since Galileo, had, like Pauli, made key contributions to quantum theory. He was instantly captivated by the Austrian physicist's idea, not simply because it solved the problem of the spread of energy of beta particles, but because it also fixed another problem: that of spin.

Physicists had discovered that subatomic particles behave as if they are spinning, even though they are not. Like everything else in the submicroscopic quantum realm, spin comes in indivisible chunks, or 'quanta'. Since a spinning charge acts like a tiny magnet, it is possible to deduce the spin of a particle from the way in which it is deflected by a magnetic field. The proton, neutron and electron all turn out to have a spin of ½. (For historic reasons, the smallest chunk is half of a particular value.)[15] In recognition of the behaviour of particles with 'half-integer spin', an idea that was elucidated principally by Fermi, they are known as 'fermions'.

Spin, like electric charge and momentum, is one of those quantities that never changes, or is conserved.* However, if a neutron (spin ½) changes into a proton (spin ½) and an electron (spin ½), the final spins add up to either 1 – if the proton and electron spin the same way – or 0 if they spin in opposite directions and their spins cancel each other out. Neither of these is the spin ½ of the initial neutron. However, Pauli, in his letter to the Tübingen meeting, had not only proposed that the neutrino has no electric charge, very little mass and that it interacts with normal matter very rarely – he had postulated that it has a spin of ½. This made it possible for the spins of the proton, electron and neutrino (½ + ½ – ½ = ½) to equal the spin of the initial neutron (½).

Never before in the history of physics had anyone predicted the existence of a new entity that solved so many problems simultaneously and whose characteristics – spin, electric charge, mass and ability to penetrate matter – were so precisely pinned down by experimental observations. It caught Fermi's imagination to such an extent that, after the October 1931 meeting in Rome, he was spurred to develop a revolutionary theory of beta decay.[16]

In the couple of years it took for Fermi to incubate his ideas, two new subatomic particles came to light, as mentioned earlier. In August 1932, Carl Anderson, studying 'cosmic rays' at the California Institute of Technology, found the first particle of 'antimatter' – a positively charged twin of the electron, which he christened the 'positron'.† And in January 1932, James Chadwick

---

* This is not strictly true. It is 'angular momentum' that is conserved. Spin is simply intrinsic angular momentum.
† Cosmic rays are high-speed atomic nuclei, mostly protons, from space. Low-energy ones come from the Sun, while high-energy ones

at Cambridge University discovered a second constituent of the nucleus, identical in mass to the positively charged proton but with no electric charge. It was the discovery of the 'neutron' that caused Fermi to suggest a new name for Pauli's hypothetical particle, neutrino being Italian for 'little neutral one'.

Fermi's theory of beta decay, when it was published in 1934, was a triumph. It required the existence of a third fundamental force of nature, in addition to the well-known gravitational and electromagnetic forces. The new 'interaction', which Fermi christened 'the weak force', operated only over a very short range within the atomic nucleus, which was why nobody had noticed it before. It acted to change a neutron in a nucleus into a proton and simultaneously create an electron and an antineutrino.

Fermi's theory also permitted the reverse process, in which a proton captured a neutrino, causing it to change into a neutron and emit a positron. (In fact, this is the process that creates a neutrino; beta decay creates an antineutrino, which was what Pauli was actually describing.) The physicists Hans Bethe and Rudolf Peierls immediately pointed out that such 'inverse beta decay' would, in theory, permit a neutrino flying through space to be stopped by matter and to therefore be detected, though this would happen extremely rarely.

Fermi did not call the new interaction the weak force for nothing. It was about ten trillion times weaker than the electromagnetic force that holds together the atoms in our bodies. It was so weak, in fact, that the chance of a neutrino being stopped by a proton in an atomic nucleus was calculated to be close to

---

probably come from supernovae. The origin of ultra-high-energy cosmic rays, particles millions of times more energetic than anything we can currently produce on Earth, is one of the great unsolved puzzles of astronomy.

zero.[17] Whereas Pauli had thought a neutrino might be halted by a piece of lead about ten centimetres thick, according to Fermi's theory it would require a layer of lead many light years thick.*†[18] As the American novelist Michael Chabon would later observe, 'Eight solid light years of lead . . . is the thickness of that metal in which you would need to encase yourself if you wanted to keep from being touched by neutrinos. I guess the little fuckers are everywhere.'[19]

Despite Fermi's theory of beta decay bolstering the case for the neutrino, many remained sceptical of its existence. And who could honestly blame them? As Nobel Prize-winning American physicist Leon Lederman would one day observe, 'Neutrinos . . . win the minimalist contest: zero charge, zero radius, and very possibly zero mass.'[20]

One of the sceptics was the English astronomer Arthur Eddington. 'Just now nuclear physicists are writing a great deal about hypothetical particles called neutrinos supposed to account for certain peculiar facts observed in beta-ray disintegration,' he said. 'We can perhaps best describe the neutrinos as little bits of spin-energy that have got detached. I am not much impressed by the neutrino theory.'

Eddington stopped short of saying that he did not believe in neutrinos. 'I have to reflect that a physicist may be an artist, and you never know where you are with artists.' If neutrinos did

---

* In quantum theory, fundamental forces are caused by the exchange of force-carrying particles. A weak force is therefore one in which force-carrying particles are exchanged rarely, and a strong force is one in which they are exchanged frequently. This is why neutrinos, which are subject to weak force, interact with other particles so rarely.
† A light year is the distance light travels in a vacuum in a year. It is roughly equal to ten trillion kilometres.

exist, Eddington recognised the problem of proving it, but even here he was cautious. 'Dare I say that experimental physicists will not have sufficient ingenuity to make neutrinos? Whatever I may think, I am not going to be lured into a wager against the skill of experimenters,' he said. 'If they succeed in making neutrinos, perhaps even in developing industrial applications of them, I suppose I shall have to believe – though I may feel that they have not been playing quite fair.'[21]

The undetectability of neutrinos was a major concern even to those who believed in their existence. The irony is that Pauli, a man who so feared loneliness, had postulated the existence of the loneliest entity in creation – a particle so mind-bogglingly antisocial that it interacts with hardly anything in the universe. 'I have done a terrible thing,' he said. 'I have postulated a particle that cannot be detected.' Leading physicists were in agreement that finding the neutrino would be impossible, and Pauli himself bet a case of champagne that nobody would ever catch one.

Los Alamos, New Mexico, November 1955

Frederick Reines had been doing the impossible for more than a decade. When he joined the Manhattan Project in 1944, it had seemed impossible that they would be able to create a runaway nuclear chain reaction, releasing a million times more energy, pound for pound, than dynamite. But they achieved that feat at Alamagordo on 16 July 1945. 'I have become Death, the destroyer of worlds,' Robert Oppenheimer, director of the Manhattan Project, had quoted from the Bhagavadgita, as they watched a mushroom cloud rise into the dawn sky above the New Mexico desert.

Later, it had seemed impossible that they could create the 'super', a device that used an atomic bomb as a trigger and unleashed the energy of the Sun itself. But they achieved that feat too, with the detonation of the hydrogen bomb in Enewetak Atoll on 1 November 1952.

They had always been faced with impossible challenges, but they had met them head-on, and triumphed. For the test of a boosted atomic bomb in 1951, for instance, they had known that their electronics would be fried when the intense flash of gamma rays from the explosion generated a huge surge of electricity in the signal cables running from the bomb tower to the instrumentation bunker. The only thing providing shielding on the scale they required was the island on which they were testing the bomb, so they simply dug up one side of the island and piled it on top of the other.[22]

The impossible challenges of the bomb tests had instilled in all of them a 'can-do' spirit and a tendency to 'think big'. It was exactly this mindset that had led Reines to seriously consider the impossible challenge of detecting the neutrinos from the explosion of a nuclear bomb.

In 1951, he had returned to the US from a series of successful bomb tests on Enewetak Atoll. Tired and jaded after six gruelling years with the weapons programme, he was in desperate need of a break. He asked the leader of the theoretical division at Los Alamos for time off from his duties to think about fundamental physics, and Carson Mark, who was an enlightened man, granted him his request. Reines was given a bare office, where he sat staring at a blank pad of paper for several months. He asked himself what he wanted to do with his life, and for a long time he did not know. But then he thought of the neutrino.

At Los Alamos, Reines had served on the 'Bomb-Test Steering and Liaison Group'. On occasion, it had tossed around the wild idea of piggy-backing physics experiments on nuclear tests and using the intense burst of heat radiation, gamma rays and neutrons to study fundamental phenomena. Reines knew that a nuclear fireball generated one additional type of radiation. When a nucleus of uranium or plutonium 'fissions', it creates two unstable 'daughter' nuclei. Each nucleus, in its desperate quest for stability, undergoes on average six beta decays, every time spitting out an antineutrino. As a result, a nuclear explosion creates an intense burst of antineutrinos.

The chance of detecting one antineutrino was impossibly low, but if there were vast numbers of them, Reines reasoned, the odds of ensnaring one would be hugely improved.

One day, in the summer of 1951, Reines heard that Enrico Fermi himself was visiting Los Alamos and was installed in an office just down the corridor. Since creating the theory of beta decay in Rome in the early 1930s, Fermi had won the 1938 Nobel Prize in Physics and fled Mussolini's fascist dictatorship for America. On 2 December 1942, he had changed the course of history: in a crude 'pile' of uranium and graphite on a squash court under the West Stands of the University of Chicago's Stagg Field, he had unleashed the stupendous energy of the atomic nucleus in the world's first sustained nuclear chain reaction.

Reines knocked nervously on Fermi's door. When he told him of his plan to detect neutrinos in the blast of a nuclear explosion, Fermi, to his surprise, did not dismiss the idea out of hand and agreed that a nuclear explosion offered the best chance of detecting the elusive particles.

A neutrino had very little chance of being stopped by a proton in an atom; the way to boost the chance was to put together lots of atoms. Reines estimated that, with a detector mass of about a tonne, it would be possible to detect a handful of neutrinos, but neither he nor Fermi had any idea of how to go about it.

The fact that Fermi had not ridiculed his idea gave Reines confidence that detecting the neutrino was possible, but the problem was that he was just one man with an obsession. That changed when he flew to a meeting in Princeton, New Jersey. The plane had engine trouble and was forced to land in Kansas City. Travelling with him from New Mexico was a physicist called Clyde Cowan, who had worked with the British on radar in the Second World War and had arrived at Los Alamos in 1949. Although Reines and Cowan had been part of the same American bomb teams, they had never had a chance to talk properly. Now, as they strolled through the streets of Kansas City while waiting for their plane to be fixed, they hit it off.

Their conversation quickly turned to fundamental physics and the question: What was the hardest experiment in the world? Both men agreed it was the detection of the neutrino. The fact that everyone thought it was impossible made it appealing to them, and they imagined the buzz of achieving something that everyone said could never be done. There and then, the two men decided to work together on detecting neutrinos. Reines had found his partner in crime.

Back at Los Alamos, there was much enthusiasm for the venture, which resulted in the creation of a neutrino group in late 1951. Since the neutrino was a fleeting ghost that barely haunted the world of physical reality, the quest to detect it was christened 'Project Poltergeist'.

The way to snare a neutrino, as Bethe and Peierls had already realised, was via inverse beta decay. On rare occasions, an anti-neutrino interacted with a proton, creating a neutron and a positron in the process. The positron would quickly run into an electron, since electrons are ubiquitous in matter, and 'annihilate' with it. There would emerge two high-energy photons, or gamma rays, flying away in opposite directions. It was these gammas – proxies for the antineutrino – that Reines and Cowan intended to detect; proving the existence of the antiparticle would automatically prove the existence of the particle, in this case the neutrino.

A year earlier, in 1950, several teams had discovered transparent liquids that emit flashes of light when a charged subatomic particle or gamma ray flies through them. The light flashes from such 'liquid scintillators' were weak, but could be boosted by placing 'photomultipliers' all around the scintillator, which converted the light into a measurable electrical signal.

The neutrino detector envisioned by Reines and Cowan would incorporate tanks of liquid scintillator and a bath of water. The protons in the water would provide a large number of targets for the antineutrinos. The pair of gamma rays created in the interaction of an antineutrino and a proton would fly out through tanks of liquid scintillator on either side of the water bath, and photomultipliers arranged around each tank would detect them.

The experiment was little more than a dull piece of plumbing, but the place where Reines and Cowan intended to locate it was anything but dull. And it was here that the fearless thinking and sheer chutzpah of the two physicists came to the fore. An atomic bomb creates a blisteringly hot fireball capable of erasing a city, and Reines and Cowan planned to place their detector a mere fifty metres from the centre of such an inferno.

Nothing in the open could survive such a blast, but Reines and Cowan envisaged placing the neutrino detector in a vertical shaft ten feet in diameter and 150 feet deep. They would pump the air out of the shaft and, at the very instant the bomb went off, would let the detector drop. During its two-second fall, not only would it be shielded from the ferocity of the fireball by the surrounding ground, but because it was in free fall, it would be protected from the potentially catastrophic shockwave thundering through the earth. At the bottom of the shaft, the fall of the apparatus would be cushioned by a thick bed of foam rubber and feathers. Reines and Cowan intended to retrieve the detector several days later, when radiation levels would be low enough to risk a quick in-and-out foray.

The extraordinary plan was granted approval by Norris Bradbury, director of Los Alamos, and work even started on digging the 150-foot-deep shaft to house the detector at the bomb test site in Nevada. But then, in the autumn of 1952, Jerome Kellogg, leader of the Los Alamos Physics Division, asked Reines and Cowan whether it might be possible to carry out the experiment with a nuclear reactor rather than a nuclear bomb. At first sight, it did not look promising – a nuclear reactor was a weaker source of neutrinos than a nuclear explosion by a factor of a thousand. However, when Reines and Cowan investigated it in detail, they were surprised to find that such a neutrino experiment was indeed possible.

When an antineutrino hit a proton, it created not only a positron, which could be detected by the two gamma rays created by its annihilation, but a neutron. The key thing, Reines and Cowan realised, was to detect the neutron as well as the positron. The neutron could be detected by taking a substance like cadmium that acted like a neutron-sponge and adding it to the liquid scintillator. Each neutron would ricochet from nucleus to nucleus, before burying itself in a cadmium nucleus after about five millionths of a second and shedding its surplus energy as a gamma ray.

The electronics connected to the photomultipliers could be arranged to register a response only to a signal consisting of two gamma rays (from the annihilation of the positron) followed by a single gamma ray (from the capture of the neutron). This 'delayed coincidence' signal was so distinctive that it was unlikely to be mimicked by any other particle process going on in the detector. The ability to reject other confusing signals would make the detection of neutrinos at a nuclear reactor possible, even though it was a far weaker source of the particles than a nuclear explosion.

A nuclear reactor had other advantages over a nuclear explosion. Rather than providing an ultra-brief window of a second or two in which to detect neutrinos, it could be monitored continuously for weeks, months or even years. Furthermore, there was no risk of the experiment being incinerated or of harm to anyone who retrieved it from a radiation-scarred landscape.[23]

In the early spring of 1953, Reines and Cowan's team loaded their vehicles with a 300-litre neutrino detector, barrels of liquid scintillator and racks of electronics, and headed for the plutonium-producing reactor at the Hanford Engineer Works in Washington state. America's newest and largest reactor was expected to generate the largest flux of antineutrinos. If it had been possible to see neutrinos with the naked eye, the reactor would have glowed like a second sun.

But at Hanford, Project Poltergeist hit a show-stopper. The neutrons the team was looking for turned out to be not the only neutrons in town; it became clear that there were others coming from fissioning nuclei in the reactor core. To absorb them and stop them reaching the detector, the team built a thick wall of paraffin, borax and lead around their experiment. It worked, but then they encountered another problem: the neutrons from the reactor core were not the only source of a signal that mimicked the signal they were looking for. There was another source, and it came from space.

Cosmic rays are high-energy nuclei created by exploding stars and other violent cosmic events. At the top of the Earth's atmosphere, they slam into nuclei of atoms and create 'secondary' particles, which shower down through the atmosphere like a fine rain. The most penetrating of all the particles are 'muons', a form of heavy electron. Cosmic ray muons slammed into nuclei in the shield that Reines' team had built around their experiment, creating sprays of neutrons. Unfortunately, these neutrons were ten times more abundant than those expected from the neutrons produced by neutrinos. 'The lesson of our work was clear: it is easy to shield out the noise men make, but impossible to shut out the cosmos,' said Cowan. 'We felt we had the neutrino by the coattails, but our evidence would not stand up in court.'

Reines and Cowan were disappointed, but not defeated. They at least knew that the technology they were using worked; all they needed was a nuclear reactor that was better shielded from the confusing signals from cosmic rays. Finally, they found one in P Reactor at the Savannah River Plant. By virtue of the fact that it was buried twelve metres deep in the ground, it was perfectly shielded from the menace from space. In November 1955, Project Poltergeist moved to South Carolina.

Savannah River, South Carolina, 14 June 1956

Reines drove past the sign a joker on the team had put up –
'DANGER, DO NOT STAND CLOSE TO FENCE
– HIGH NEUTRINO FLUX' – and parked next to the
empty truck that had brought a load of wet sawdust to the
site.[24] The control room, with its humming generator, was in
a trailer dwarfed by the concrete hulk of the reactor. Cables
snaked across the ground, carrying the electrical signals up
from the scintillator tanks twelve metres below and creating
a trip-hazard that Reines had to be careful to avoid. Inside,
Cowan was sitting in front of a wall of oscilloscopes, switches
and racks of glowing vacuum tubes, monitoring the output
from the detector.

With its team of almost a dozen and a mountain of accompanying equipment, the ten-tonne detector was the biggest
physics experiment on the planet. Nobody before had dared
carry out anything as complex, but nobody else had learnt their
craft while testing the weapons of Armageddon or had at their
disposal the financial resources, machine shops and technology of Los Alamos. This was big science. It was a vision of the
future: one day much of physics would be done like this, in lab-
oratories that spanned national borders, employed thousands
of researchers and cost tens of billions of dollars.

The final design for the Project Poltergeist apparatus was
a kind of double-decker sandwich. Two layers of water with
cadmium chloride added to it acted as the neutrino target,
and these were interleaved with three layers of liquid scintillator. Positrons produced by neutrinos interacting with protons in the water would be detected almost immediately via
back-to-back gamma rays in the adjacent scintillator tanks,

and neutrons produced by the same neutrinos would reveal themselves five microseconds later via another burst of gamma rays in the same tanks.

Project Poltergeist had been running for 1,371 hours. Not only was the gamma ray signal it had measured four times bigger than the background level, it was five times bigger when the reactor was switched on than when it was switched off. Every hour they detected three neutrinos.

The possibility remained, however, that neutrons from nuclear fissions in the reactor were penetrating the eleven metres of concrete shielding around the reactor and creating spurious gamma rays in their experiment. So overnight, while Reines had slept, the rest of the team had piled bags of wet sawdust against the wall of the reactor. Their material of choice – a tribute to the cuisine of South Carolina – had actually been black-eyed peas, but wet sawdust was easier and cheaper to obtain in the required quantities.[25]

If some of the gamma rays they were detecting were coming from neutrons produced by the reactor, the extra shielding of the wet sawdust should have stopped them, reducing the signal by a factor of ten. 'Any change in the signal?' asked Reines. Cowan, looking up from an oscilloscope, grinned. 'No change.' It was exactly what Reines had wanted to hear.

Outside the trailer, the whole team had assembled: Richard Jones and Forrest Rice, who had installed the detectors and lead shielding; F. B. Harrison, the expert in large liquid scintillators; Austin McGuire, who had designed the tank farm containing the scintillator; Herald Kruse, who had been responsible for interpreting the oscilloscope traces; and Martin Warren, the gopher. All of them looked exhausted, but they were euphoric and pumped each other's hands, slapping each other on the back. They had overcome the final hurdle and achieved the impossible. After five years of sweat and struggle, they had detected the elusive neutrino.[26]

There remained only two things to do: send a telegram and then pack up their equipment and drive back to Los Alamos.

o o o

Pauli received the telegram on 14 June 1956: 'We are happy to inform you that we have definitely detected neutrinos from fission fragments by observing the inverse beta decay of protons . . . Frederick Reines, Clyde Cowan.'

The next day, Pauli replied from the Swiss Federal Institute of Technology in Zurich: 'Frederick REINES, and Clyde COWAN, Box 1663, LOS ALAMOS, New Mexico. Thanks for message. Everything comes to him who knows how to wait. Pauli.'

With the proof of the neutrino that he had predicted a quarter of a century earlier, Pauli well and truly joined the ranks of the magicians. Who in their most extravagantly wild dreams would have imagined something as insubstantial, ghostlike and downright weird as the neutrino? Pauli had predicted it for the sole reason that it was what the mathematical logic was telling him. The neutrino simply had to exist because, without it, radioactive beta decay made no sense at all.

Pauli announced the discovery of the neutrino at a symposium at CERN, the European laboratory for particle physics near Geneva, the week after he received Reines' telegram. Reines, in his Nobel Prize acceptance speech in 1995, would recount that Pauli celebrated with a case of champagne.[27] That certainly made a good story since Pauli had, of course, bet a case of champagne that the neutrino would never be detected, though sadly it was not true.[28]

Reines and Cowan, for their part, were rather more sober. Standing outside P Reactor in the South Carolina sunshine, they and their team celebrated their success not with flutes of champagne but with paper cups of Coca-Cola.[29]

o   o   o

For the neutrino, it was only the beginning of the story. Hold up your hand. About one hundred billion neutrinos pass through your thumbnail every second. Eight and a half minutes ago they were in the heart of the Sun. Solar neutrinos are produced in prodigious quantities by sunlight-generating nuclear reactions.

Remarkably, Reines and Cowan's team was not the only one with the temerity to attempt the impossible feat of detecting the neutrino; it was not even the only one to do it at the Savannah River Plant. In 1954, a team led by Raymond Davis, the American chemist and physicist, had installed a detector filled with 3,800 litres of cleaning fluid – carbon tetrachloride – in the basement of one of the nuclear reactors. The idea behind the detector had been suggested by Bruno Pontecorvo, a one-time colleague of Enrico Fermi's who had defected to the Soviet Union. Occasionally, a neutrino would interact with a chlorine nucleus in the cleaning fluid, turning it into a nucleus of argon, a gas which could be easily separated. The amount collected would correspond to the number of neutrinos detected.

Unfortunately, Davis had suffered similar problems to Reines and Cowan at Hanford. His detector was not sufficiently shielded from the confusing effect of cosmic rays and so he lost out in the race to detect the neutrino. But he was nothing if not persistent. In the mid-1960s, he located a detector of 400,000 litres of cleaning fluid 1.5 kilometres underground in the Homestake gold mine in Lead, South Dakota. His aim was to detect neutrinos from the core of the Sun, and incredibly, he succeeded, becoming the first person to see into the heart of a star.

But there was a problem. Unexpectedly, Davis registered only between one third and a half of the neutrinos which were predicted by the theory of energy generation in the Sun. Was there something wrong with our understanding of the Sun, or with our understanding of neutrinos?

Davis's conundrum triggered a wave of other experiments to check his anomalous result, and he is credited with giving birth to the field of 'neutrino astronomy'. The belief of pretty much everyone was that Davis was wrong. Contrary to expectations, however, the new experiments confirmed that there was indeed a shortfall in the number of neutrinos coming from the Sun.[30]

The 'solar neutrino puzzle' had an extraordinary solution, which was eventually confirmed by the Sudbury Neutrino Observatory in Ontario, Canada: there are three types of neutrino. These are the electron neutrino; the muon neutrino, discovered in 1962 at Brookhaven, New York; and the tau neutrino, discovered in 2000 at Fermilab near Chicago. Nobody knows why nature has chosen to triplicate its neutrinos – along with all its other basic building blocks, the quarks – but crucially, the observatory could detect all three types. What it showed, in 2006, was that there was no shortfall in neutrinos, as long as the numbers of the three types were added together.

As early as 1957, Pontecorvo had suggested neutrinos might come in different types, or 'flavours', and that, flying through space on their way from the Sun to the Earth, they might morph from one type to another. Imagine a dog walking along a street and changing into a cat after one hundred metres, a rabbit after another one hundred metres and back into a dog after a further one hundred metres. Say, for some weird reason, your eyes can only spot dogs: you would see them only a third of the time. So it was with neutrinos. Davis's experiment was sensitive only to electron neutrinos, but the neutrinos arriving at his detector in the Homestake mine were in the guise of electron neutrinos only a third of the time.[31]

Such neutrino 'oscillations' have implications for the mass of the neutrino, which many had assumed was zero. According to Einstein's special theory of relativity, only a massless particle like the photon can travel at the ultimate cosmic speed limit – the speed of light – and, for such a particle, relativity predicts that time slows to a standstill. The photon cannot therefore change since change is something that can only happen in time. However, the neutrino emphatically does change, oscillating between its three flavours. The implication is that it must travel slower than the speed of light and therefore have a mass.[32]

The mass of the neutrino is, not surprisingly, hard to measure. It appears to be at least 100,000 times smaller than that of the electron, which was formerly the lightest known subatomic particle. This suggests that neutrinos acquire their mass in a different way to all the other fundamental particles, which get theirs by interacting with the 'Higgs field' (see chapter 'The god of small things'). The Higgs is a key component of the Standard Model of particle physics, a quantum description of nature's three non-gravitational forces. Although very successful, the Standard Model fails to predict the masses of the fundamental particles or the relative strengths of the fundamental forces and is widely believed to be an approximation of a deeper, more satisfactory theory. The hope among physicists is that if they can understand how the neutrino gets its mass, they might gain important clues about this elusive 'theory of everything'.

Despite the fact that the mass of the neutrino is extremely small, neutrinos could still have important consequences for the universe. Prodigious quantities of them flood out of the Sun, not to mention every other star in the galaxy, and they were also created in uncountable numbers by processes in the Big Bang that created the universe 13.82 billion years ago.[33] Neutrinos are the most elusive entities in nature, as close to nothing as anything we know of and apparently spectators rather than participants in the life of the cosmos. However, they turn out to be the second most common particles in nature, after photons. In terms of sheer numbers, we live in a neutrino-and-photon universe.

So even though neutrinos have ultra-tiny masses, they could still make up a significant fraction of the mass of the universe. In fact, if there exists an as-yet-undiscovered massive neutrino, neutrinos could be a component of the universe's mysterious dark matter, which is known to outweigh the visible stars and galaxies by a factor of about six.[34]

But this is not the only way in which neutrinos could be the key to the universe. Experiments showing differences between the rates of creation and destruction of neutrinos and anti-neutrinos hint at a fundamental asymmetry between matter and antimatter. It may one day explain one of the biggest mysteries of the universe: why we live in a universe of matter that contains virtually no antimatter.[35]

Once Reines' team detected the neutrino in 1956, he was far from finished. In 1987, he was a member of one of two teams that detected a total of nineteen neutrinos coming from another star. Supernova 1987A marked the detonation of a massive star in the Large Magellanic Cloud, a satellite galaxy of the Milky Way. It was the first supernova seen in our galaxy for four hundred years.

When a massive star reaches the end of its life, it runs out of fuel to generate the internal heat necessary to oppose the gravity trying to crush it. As the core shrinks catastrophically, heating up to ferocious temperatures, the elements built up by nuclear reactions over the star's lifetime come apart into protons, neutrons and electrons. Electrons are squeezed into protons to create a superdense ball known as a 'neutron core', in the process unleashing a tsunami of neutrinos. In the case of Supernova 1987A, this amounted to 10^{58} – or ten billion trillion trillion trillion trillion – neutrinos. Although a supernova can shine as bright as a galaxy of 100 billion stars, it turns out that a mere 1 per cent of its energy is emitted in the form of light; 99 per cent consists of neutrinos.

It is the neutrinos flooding out of the star that turn the implosion of the core into a supernova explosion, blowing the exterior envelope of the star into space. This contains elements which enrich interstellar gas clouds, destined to become stellar nurseries when they fragment into the new generations of stars. Without neutrinos, the elements essential for life would remain locked up inside stars. 'Why does nature need them? What use are they?' asks the English physicist Frank Close.[36] The remarkable truth of Pauli's 'impossible particle' is that it is more critical to the universe than anyone could possibly have imagined. Without it, you would not be reading these words; in fact, you would not even have been born.

6

The day without a yesterday

The radiation left over from the Big Bang is the
same as that in our microwave oven but very much
less powerful. It would heat your pizza only to
minus 270.4°C – not much good for defrosting the
pizza, let alone cooking it!

STEPHEN HAWKING

The elements were cooked in less time than it takes
to cook a dish of duck and roast potatoes.

GEORGE GAMOW

Holmdel, New Jersey, Spring 1965

For the best part of a year, they had faced nothing but delay and
frustration. For the best part of a year, a stubborn hiss of radio
static had prevented them from doing the slightest bit of astronomy. But now, as they put on overalls and boots and climbed
into the gaping mouth of the twenty-foot 'horn' with their stiff
brooms, Arno Penzias and Robert Wilson were convinced that
their nightmare would soon be over.

The horn, a giant metal funnel the size of a railway car-
riage, stood on Crawford Hill, a wooded knoll near Holmdel,
New Jersey. It belonged to Bell Labs, part of AT&T, the giant
American phone company, and it had been built in 1959 to test
the feasibility of beaming phone and TV signals around the
Earth via a ring of communications satellites.

The idea of such satellites had been proposed by the British science-fiction writer Arthur C. Clarke.[1] In an article in the October 1945 issue of *Wireless World*, he had pointed out that the further a body is from the Earth, the weaker it is gripped by the planet's gravity and so the more sluggishly it orbits. At a special distance – 35,787 kilometres from the Earth's centre – the body circles so slowly that it orbits the planet once every twenty-four hours. Observed from the ground, such a satellite would appear to hang motionless in the sky.

Clarke's idea was to have three communications satellites equally spaced around this 'geosynchronous orbit'. Sending a phone conversation from England to, say, Australia would then involve broadcasting a radio signal from a transmitter in England to the nearest satellite, which would relay it to the next satellite and the next one, before beaming it back down to a receiver in Australia.

In 1945, the idea of a planet girdled by communications satellites was the wildest of science fiction. But Clarke, while serving as a radar technician with the Royal Air Force, had vivid memories of the Nazi bombardment of London by V2 ballistic missiles, and had realised that such rockets could just as easily be fired straight upwards as at a distant city. He was not alone in thinking this. On 24 October 1946, a V2 rocket captured by the Americans and launched from the White Sands Missile Range in New Mexico escaped the atmosphere and took the very first photograph of the Earth from space.

Clarke's belief that science fiction was rapidly becoming science fact was confirmed on 4 October 1957, when the Russians launched the first satellite. Sputnik 1, a metal sphere fifty-eight centimetres in diameter that beeped incessantly as it circled the Earth, terrified the Americans, who feared the Russians dropping an H-bomb on a city like New York. It marked the birth of the 'Space Age' and kick-started the space race between the world's two superpowers. Almost immediately, AT&T and many other companies realised that they needed to get into the satellite business, and fast.

The best way to communicate with a satellite was via 'microwaves', radio waves with short wavelengths of between a few centimetres and a few tens of centimetres. The problem with microwaves is that everything glows with them – people, trees, buildings, the sky, and so on. The challenge for the AT&T engineers was to pick up a weak microwave signal from a tiny source in the sky – a satellite – amid much stronger signals coming from every other direction.

The microwave horn on Crawford Hill, the construction of which had begun in the summer of 1959, was the AT&T engineers' solution. When its twenty-square-foot opening was directed at a point-like object in the sky, microwaves from all other sources had difficulty bending their way into the horn, meaning that only microwaves from the desired source were funnelled down to the tapered end of the horn, where they were detected by a radio receiver.

The first test of the twenty-foot horn was Echo 1, a kind of Stone Age communications satellite launched by NASA in 1960. It was, in effect, a 100-foot-diameter silvered inflatable beach ball, off which radio waves from the microwave horn could be bounced and picked up (a radio horn has the ability to both transmit and receive radio waves). Hot on the heels of Echo 1 came the first modern communications satellite. Telstar did not simply passively bounce back radio waves transmitted from the ground; it boosted them in strength before retransmitting them. In 1962, it relayed the first-ever television pictures between America and Europe. Telstar caused a global sensation, and pop records were even recorded about the satellite.

By 1963, when the world had well and truly entered the age of communications satellites, AT&T no longer needed the twenty-foot horn on Crawford Hill, so decided to hand it over to some radio astronomers. This was not an altruistic act. Such astronomers were in the business of detecting ultra-weak signals in the sky, just as AT&T was, and the company reasoned that it might benefit from giving the horn to science. In fact, it was not AT&T's first venture into astronomy. In the 1930s, the company had employed Karl Jansky to identify sources of radio interference which were playing havoc with wireless reception. By picking up radio waves from the Sun and a mysterious source at the centre of the Milky Way that later turned out to be a 'supermassive' black hole, Jansky earned the title of the 'father of radio astronomy', and the unit of radio power is called a Jansky in his honour.

Arno Penzias, a dynamic thirty-one-year-old New Yorker whose family were refugees from Nazi Germany, had arrived at Holmdel in 1962. Robert Wilson, a taciturn twenty-eight-year-old from Caltech in Pasadena, came in early 1964, and in the summer of that year, the pair teamed up.

Wilson had the suspicion that the Milky Way, our galaxy, which is shaped rather like a CD, might be embedded in a spherical halo of extremely cold hydrogen gas left over from the galaxy's formation. If so, the gas would be glowing with very faint radio waves, and the horn, because of its ability to reject spurious radio waves from its surroundings, had a unique ability to pick up such a signal.

However, a modification was necessary because the faint signal Wilson intended to look for would likely be drowned out by the radio waves emitted by the sky. Radio astronomers usually overcome this problem by rapidly switching their telescope between an astronomical source – a star or a galaxy – and a neighbouring patch of sky. By subtracting one signal from the other, they are able to neatly remove the emission from the sky.[2] This would not work for observing the 'galactic halo' because we are inside the Milky Way; since the halo fills the sky, it is impossible to point away from it.

The solution hit upon by Penzias and Wilson was to compare the galactic halo with an artificial source of radio waves. Penzias built such a source, which he cooled with liquid helium to four degrees above absolute zero. He housed this 'cold load' in the shed strapped to the tapered end of the horn, which also contained the radio receiver.

Before doing any actual astronomy, Penzias and Wilson made sure their equipment was working by observing at a frequency where they expected there to be no radio glow from the galactic halo. It appeared daft but there was method in their madness: if the signal detected when they pointed the horn at a patch of empty sky was precisely zero, they would know their equipment was working as expected.

However, when they carried out the test, things did not go as planned. The signal they recorded was not zero; instead, there was a residual hiss of radio static. It was what would be emitted by a body at about −270 degrees Celsius, or three degrees above absolute zero.

At first, the two astronomers thought the hiss was coming from New York, which was just over the horizon from Holmdel, but when they pointed the horn away from the city, the static remained. Their next thought was that it might be coming from a source in the solar system – both the Sun and

Jupiter broadcast radio waves – but as the months wore on and the Earth travelled around the Sun in its orbit, the static did not change. The astronomers next wondered whether the source might be a high-altitude explosion of a hydrogen bomb. On 9 July 1962, 'Starfish Prime' injected high-energy electrons into the Van Allen Belts, recently discovered regions of Earth's magnetic field that trapped charged particles from the Sun. Spiralling around the magnetic field 'lines', such electrons would be expected to emit radio waves; such an effect would obviously decline with time, but the static did not.

Finally, Penzias and Wilson's gaze settled on two pigeons that had made a nest inside the horn. At first sight, it did not appear to be a good place to make a home; they had to remake their nest every time the horn was turned to point in a new direction. However, the winters in New Jersey are cold, and the tapered end of the horn, where they had nested, was next to the refrigerator that cooled the electronics of the radio receiver. Anyone who has been around the back of a refrigerator knows it is warm there; the pigeons had chosen a cosy spot to raise their family. In doing so, however, they had coated the interior of the horn with what Penzias and Wilson referred to as a 'white dielectric material'. To everyone else, it was pigeon shit, which, like everything else, glows with radio waves. Penzias and Wilson exchanged glances. Could they at last have found the source of the annoying hiss of radio static that had prevented them from doing any astronomy for so many months?

At a local hardware store, they bought a baited 'Havahart' trap. When a bird stepped on a finely balanced plate, a gate behind it dropped, trapping the bird. With the aid of the trap, the two astronomers caught the pigeons and posted them (!) in the company mail to another AT&T site at Whippany, New Jersey.[3] With the pigeons gone, Penzias and Wilson climbed into the mouth of the horn to put in an hour of hard work with stiff brooms. For good measure, they stuck aluminium tape over the rivets holding together the horn's metal sheets, just in case they were contributing to the radio hiss.

Back down on the ground, they changed back into their everyday clothes, full of hope that their problem was finally solved and that they would at long last be able to do some real astronomy. As the sixteen-tonne horn turned slowly on its axes, Penzias and Wilson's eyes were glued to a pen-recorder that traced a jittery straight line across a paper roll. The horn's opening arrived at the point where it was looking back at the sky. And the trace jumped.

The radio static was still there! Penzias and Wilson shook their heads in dismay. What in the world could it be?

Washington DC, Summer 1948

Ralph Alpher and Robert Herman stood for a while and admired their calculations. Written on the blackboard were the details they had painstakingly worked out during an evening of brainstorming. If they were right, the proof that the universe had been born rather than existed forever was literally fizzing in the air all around them, and it had an unmistakable signature.

The smoke from George Gamow's cigarette was still hanging in the air. 'Write it up, you two! Write it up!' Gamow had ordered when he saw their calculations. Then he had left, firing off ideas like firecrackers, as was his way. By now, he was no doubt onto something else: galaxy formation, quantum theory, analogies to use in his popular series of 'Mr Tompkins' books

– who knew what? The visits of Alpher's larger-than-life supervisor were like drive-by shootings, leaving him and Herman stunned and overwhelmed. But to give him his credit, it was Gamow who had come up with the idea that set them on their path to discovery. Though he exasperated them – largely because his practical jokes and drunkenness made other physicists see him as more fly-by-night dilettante than serious scientist – they loved him dearly.[4]

Gamow had defected from Stalin's Russia in 1933 with his wife, fellow physicist Lyubov Vokhminzeva. Unable to find a permanent academic post in Europe, he had headed for the US the following year, where he had ended up with a professorship at the George Washington University in Washington DC; this was where Alpher, who was studying at night while holding down a day job working on the theory of guided missiles, had become aware of him. Gamow was loud, enthusiastic, irreverent and larger than life in every way. He may not have been highly regarded by the physics community, but he knew all the greats personally – Albert Einstein, Niels Bohr, Werner Heisenberg – and he had made a major contribution to physics by being the first person to apply quantum theory to the nucleus of the atom, in the process explaining the long-standing mystery of radioactive 'alpha decay'.

Alpher plucked up the courage to ask whether Gamow would take him on as a doctoral student, even though he was working at Johns Hopkins University in Baltimore, and Gamow said yes. It was only later that Alpher bumped into Robert Herman, a postdoctoral student who had an office a few doors down the corridor. Herman stopped by to introduce himself, and when Alpher told him about the calculations he was working on, he was instantly hooked.

The calculations had been triggered by Gamow, who had been thinking about the origin of the chemical 'elements'. As pointed out earlier, by the 1940s it had become apparent that all ninety-two of the naturally occurring elements – from hydrogen, the lightest, to uranium, the heaviest – had not been put in the universe on day one by a Creator but had instead been *made*. The clue was in the correlation between the abundance of the elements and how strongly their nuclei were bound together. This was a powerful hint that nuclear reactions had been involved in the creation of the elements – that the universe started off with hydrogen, the simplest element, and the rest had subsequently been assembled from this basic nuclear Lego brick.

The problem with building up the elements like this was that all nuclei carry a positive electric charge; since like charges repel each other, this means they have a powerful aversion to each other. The only way for them to get close enough to stick together is for them to slam into each other at high speed, which is synonymous with high temperature. A temperature of many billions of degrees is required. But where in the universe is there such a blisteringly hot furnace?

The obvious place is deep inside stars, though Arthur Eddington had wrongly concluded that the interiors of stars were neither hot enough nor dense enough for 'nucleosynthesis'. This was the state of play in the mid-1940s, when Gamow began thinking about the problem of element-building.

Back in Russia, Gamow had been the student of Alexander Friedmann, who in 1922 was the first person to realise that Einstein's general theory of relativity implies that we live in a restless universe that has to be in motion and cannot, as Einstein himself believed, be static and unchanging. Specifically, he reasoned, the universe must be either contracting down to or expanding from a superdense state. The term 'Big Bang' would not be coined for almost three decades, but Friedmann had discovered the Big Bang solutions to Einstein's equations (Friedmann died prematurely in 1925, aged thirty-seven, which was another reason why Gamow was not tied to Russia).

In 1929, American astronomer Edwin Hubble, using the world's biggest telescope on Mount Wilson in Southern California, discovered that the universe is indeed expanding, its constituent galaxies flying apart like pieces of cosmic shrapnel, just as Friedmann had predicted. But although an explosion in the distant past was likely responsible for cosmic expansion, nobody until Gamow thought seriously about it, because it seemed so remote from everyday experience.[5]

Gamow imagined the expansion of the universe running backwards, like a movie in reverse. When everything had been squeezed into a tiny, tiny volume, it would have been hot, for the same reason that air squeezed in a bicycle pump gets hot. The Big Bang, Gamow realised, would have been a blisteringly hot fireball. Could this have been the elusive furnace in which nature's elements were forged? Gamow was not a details man – in fact, he was notorious for making errors in equations and adding things up incorrectly. He therefore gave the problem to his student, Alpher, to explore.

Neither Alpher nor Gamow knew the exact ingredients with which the universe had started, but they knew that they must have been simple. Alpher tried a number possibilities. One was a mix of protons and neutrons. Along with the proton, the neutron, discovered in 1932 by English physicist James Chadwick, is one of the two basic building blocks of all nuclei (apart from hydrogen, which contains a lone proton). By virtue of the fact that it carries no electric charge, a neutron can easily approach and stick to a nucleus. However, if too many neutrons embed themselves in a nucleus, it becomes unstable and one of its neutrons changes into a proton, a process known as 'beta decay'.

Alpher very quickly realised that, because of the rapid expansion and cooling of the Big Bang fireball, there would have been only a brief opportunity for element-building, lasting from when the universe was about one minute old to when it was about ten minutes old. After that point, the expansion would have driven nuclei so far apart and they would have been moving so slowly that their collisions would have been too infrequent and low-impact to cause them to stick to each other. Another important effect was that free neutrons decay into protons in about ten minutes, so their supply would have been rapidly depleted.

Alpher duly carried out the calculations to see what would have been the result of the orgy of 'nuclear reactions' in the Big Bang fireball. He found that the furnace would have converted about 10 per cent of all nuclei into helium, leaving the remaining 90 per cent as hydrogen. This was a remarkable result – it was exactly what was observed in today's universe.

Although this was an undoubted success that bolstered the case for the Big Bang having been the furnace in which the elements were forged, it was hard to see how any elements heavier than helium could have been made. Even if nuclear reactions had proceeded for longer than ten minutes, it would not have helped; the problem, as we know, was that nature has no stable nuclei with five or eight nucleons. Since helium has four nucleons (two protons and two neutrons), the route to building heavier nuclei – by either adding a nucleon (making a nucleus with five nucleons) or sticking two helium nuclei together (making a nucleus with eight nucleons) – is well and truly blocked.[6]

Alpher wrote a paper on his calculations, which was effectively his PhD dissertation. He co-authored it with Gamow, but his supervisor, ever the joker, decided to add Hans Bethe's name to the paper. Although Bethe had contributed nothing to the work, it meant that the list of authors now read 'Alpher, Bethe and Gamow'.[7, 8] Alpher was dismayed. As a mere student, he needed maximum credit for his work in order to secure a permanent academic post, and Gamow had muddied the waters. Bethe was a big-name physicist who had worked on the Manhattan Project and had famously figured out a chain of nuclear reactions that might power the stars on a napkin in the dining car of a train travelling between Washington DC and New York; people were bound to assume he was the driving force behind the 'cosmic nucleosynthesis' calculations. Alpher's worst fears were realised when, on turning up to defend his PhD thesis, he found himself confronted not only by Gamow and one or two of his colleagues but an audience of around 300 eager physicists.

However, nuclear reactions that built up elements were not the only consequence of a hot Big Bang. There was another one, and it was this that Alpher and Herman had been exploring, and which was the subject of the calculations Gamow had seen scrawled across the blackboard in Alpher's office.

When the universe was about a minute old and its temperature about ten billion degrees, there would have been around ten billion photons per nucleon; they would have been utterly dominant and matter a very minor constituent of the universe.[9] This prompted the question: Where did all those photons go? The answer, Gamow realised, was nowhere. Unlike the heat of the fireball of a nuclear explosion, which eventually dissipates into the surroundings, the heat of the Big Bang fireball had nowhere to go. It was bottled up in the universe, which, by definition, is all there is. Consequently, the photons of the 'afterglow' of the Big Bang must still be around us today. A quick back-of-the-envelope estimate revealed that the total energy of relic photons in any volume of space should be about the same as the total energy of starlight. Gamow concluded from this that they would be indistinguishable from starlight and that there was absolutely no chance of detecting them.

But Gamow, Alpher and Herman realised, was wrong. A crucial event in the history of the universe happened a few hundred thousand years after its birth, when the expanding fireball had cooled to about three thousand degrees. Nuclei and electrons were now flying around slowly enough that they could partner up and make the universe's first 'atoms'. This had a dramatic effect on the appearance of the universe. Whereas free electrons are very good at 'scattering', or redirecting, photons, electrons trapped in atoms are not. Consequently, before this 'epoch of last scattering', photons were forced to zigzag their way across space, like photons bouncing off water droplets in a fog. After this, the cosmic fog lifted and the universe became transparent. 'Decoupled' from matter, the Big Bang photons were able to fly unhindered in straight lines across space.

The relic photons of the Big Bang would no longer be the fiercely hot ones that began their journey 13.82 billion years ago. Greatly cooled by the expansion of the universe in the intervening aeons, they would today appear as short-wavelength radio waves, or 'microwaves'. Furthermore, they would appear to be coming evenly from every direction in the sky.

This uniform glow of microwaves was the first of two unmistakable features that Alpher and Herman realised would make the 'afterglow' of the Big Bang distinguishable from starlight. The second feature was a little more technical.

In the fireball of the Big Bang, every time a photon bounced off a free electron, the pair exchanged energy. If an electron was moving fast, the photon gained energy; if it was moving slowly, it lost energy. The collisions were frequent, and the result of huge numbers of them was that the total available energy was shared out among the photons in a very special way; very few photons ended up with low energy and very few ended up with high energy, while a lot had energies between the extremes. Such a hump-shaped energy 'spectrum' is known as a 'black body' and is particularly simple because its shape depends on only one thing: the temperature.[10] Despite the fact that the fireball of the Big Bang was expanding fast, photon–electron collisions were a lot faster, so there was time for large numbers of them before the fireball expanded appreciably. Consequently, even as the temperature plummeted, the photons retained their characteristic black-body spectrum. This spectrum was the second property that Alpher and Herman realised would make the afterglow of the Big Bang fireball distinguishable from starlight. It was necessary only to know its temperature to know everything about it.

Alpher and Herman set about the task. They worked incredibly well together, having noticed from the moment they first met that they were of one mind. It was almost as if they had a telepathic connection. Eventually, they arrived at a temperature – the number Gamow saw on the blackboard that prompted him to order them to 'write it up!' It was a chilly five kelvin (–268 degrees Celsius). Today, the afterglow of the Big Bang would appear as microwaves coming from every direction in the sky, and its spectrum would be exactly the same as that from a body at five degrees above absolute zero.

o   o   o

Alpher and Herman did as Gamow had instructed and wrote up their prediction of the 'cosmic background radiation' as a short paper. If they were right, 99.9 per cent of all the photons in the universe are tied up in the afterglow of the Big Bang and a mere 0.1 per cent in the light from the stars and galaxies. It was a remarkable claim, but might they have made some mistake? They buried their doubts and sent their paper to the British science journal *Nature*.

The paper was published on 13 November 1948, and Alpher and Herman waited eagerly for the reaction of the scientific community,[11] but it never came. Their prediction was met with a deafening silence. Not ones to give up without a fight, the two physicists mentioned their result in numerous talks over the following years. The talks were even attended by radio astronomers, whom they always made a point of buttonholing. 'Can this relic radiation from the Big Bang be detected with radio telescopes?' they asked. 'No,' came the unanimous (but incorrect) answer. And so no one embarked on a search for what, if it existed, was the single most striking feature of the universe: the afterglow of creation.

Holmdel, New Jersey, Spring 1965

It was depressing enough that the persistent, inexplicable microwave hiss was still there after Penzias and Wilson had cleaned the pigeon droppings from the twenty-foot horn. Even more depressing was the fact that the pigeons, having been sent away, returned to Crawford Hill – they were homing pigeons, after all – and had to be shot. Penzias and Wilson were not even able to console themselves with the thought that the birds had died for science; the two scientists were in the same position they had been in since teaming up the previous summer – unable to do any astronomy.

Just as they were succumbing to despair, Penzias happened to make a serendipitous phone call to a radio astronomer friend called Bernie Burke at the Department of Terrestrial Magnetism in Washington DC. It was about another matter entirely, but at the end of the conversation, Penzias could not stop himself moaning about the annoying hiss of static they were picking up at Crawford Hill. Burke sat up. He had recently been to a talk by a Princeton researcher called Jim Peebles and recalled that Peebles' boss, Bob Dicke, was supervising the building of a small radio telescope on the rooftop of the geology building at Princeton to look for microwaves that had survived from a possible hot, dense phase of the early universe.[12]

Immediately, Penzias got off the phone to Burke and phoned Princeton. At the time, Dicke was having a 'brown bag' lunch with his team in his office. There followed a short, technical exchange, involving phrases like 'microwave horn' and 'cold load'. The members of Dicke's team exchanged glances, and when Dicke put down the phone, they already suspected what followed. 'Well, boys,' he said, shaking his head. 'We've been scooped.'

The next day, Dicke's group drove over to Holmdel, only thirty miles from Princeton. After examining the twenty-foot horn, the radio receiver and the cold load, and talking briefly with Penzias and Wilson about their experimental set-up,

Dicke admitted that the game was up; the two astronomers had stumbled upon precisely what the Princeton team had been planning to look for.

Most of the light in the universe is tied up in the afterglow of the Big Bang. If you were to tune an old 'analogue' TV between the stations, 1 per cent of the 'static' on the screen would be from the Big Bang. Before being intercepted by your TV aerial, it has travelled across space for 13.82 billion years, and the last thing it touched was the fireball of the Big Bang. By discovering the cosmic background radiation, Penzias and Wilson had helped prove that the universe had not existed forever but had been born in a fireball: the Big Bang.[13]

The two teams – Penzias and Wilson, and Dicke's group – decided to announce their discovery in back-to-back papers in *The Astrophysical Journal Letters*. Ironically, Penzias and Wilson were adherents of the idea proposed in 1948 by Fred Hoyle and two colleagues that the universe has existed forever and had no hot, dense beginning. As supporters of this 'steady-state theory', they were not happy about making a claim that what they had stumbled on was evidence of the rival Big Bang theory. In their paper, they mentioned only the annoying hiss of static, which they believed was an experimental result that would hold up whatever, while leaving the speculation about the precise identity of the hiss to the accompanying paper by Dicke's team.

Two weeks before the papers were due to be published, the phone rang at Crawford Hill, and Penzias picked it up. It was Walter Sullivan, a science reporter at the *New York Times*. He had been on the trail of another story when he happened to call the offices of *The Astrophysical Journal*, and an editor had let slip that the journal was about to publish papers reporting a mysterious radio signal that was potentially from the beginning of time. Sullivan grilled Penzias about the work with the twenty-foot antenna.

Wilson's father was visiting him from Texas at the time. An habitual early riser, the next day he got up well before his son and walked to the local drugstore. When he came back, he thrust a copy of the morning's newspaper into the face of his bleary-eyed son. There, on the front page of the *New York Times*, was a picture of the twenty-foot horn, with an account of *The Astrophysical Journal Letters* papers.

o   o   o

Gamow, by now retired and living in Boulder, Colorado, read the story in the *New York Times*, but to his dismay could see no mention of his name, nor those of Alpher or Herman. It is fair to say that he awaited the publication of the papers in *The Astrophysical Journal Letters* with intense interest.

The title of Penzias and Wilson's paper was a masterclass in caution and dullness: 'A Measurement of Excess Antenna Temperature at 4,080 Megacycles per Second'.[14] Basically, all the two astronomers said was that 'Measurements of the effective zenith noise temperature of the twenty-foot horn-reflector antenna at the Crawford Hill Laboratory, Holmdel, New Jersey, at 4,080 megacycles per second have yielded a value of about 3.5 degrees higher than expected.' Nowhere in their brief paper did they mention that the radiation they had picked up might have come straight from a hot Big Bang. They merely noted, 'A possible explanation for the observed excess noise temperature is the one by Dicke, Peebles, Roll and Wilkinson in a companion letter in this issue.'

Tipped off by Gamow, Alpher and Herman had made a bee-line for their respective libraries the moment the two scientific papers came out. They could hardly believe that the Big Bang radiation had finally been found – that the stuff they had predicted on a blackboard in Washington DC seventeen years earlier was truly out there. It actually filled the universe, just as they had imagined. Both men raced to the end of the papers, and when they got there, they stood in a state of shock. Nowhere was there a mention of Alpher and Gamow's groundbreaking work on element-building in a hot Big Bang. And nowhere was there a mention of Alpher and Herman's prediction of the heat afterglow of the Big Bang. They had shown themselves to be magicians, but it looked as if nobody would ever know.

It was scarcely credible that they had been overlooked. Not only had they published the results of their hot Big Bang calculations in a series of technical articles in *Physical Review*, they had also written numerous popular accounts of their work. In fact, in 1952, Gamow had published a book for lay readers called *The Creation of the Universe*, in which he talked about the cooking of helium in a hot Big Bang and how this was connected to the temperature of the universe. In 1956, he had even aired his ideas in an article in the popular magazine *Scientific American*.

Other scientists were now taking credit for work they had done almost two decades earlier. For Gamow, Alpher and Herman, it was almost too much to bear.[15]

Holmdel, New Jersey, Autumn 1978

Wilson got the first inkling about the prize in early 1978. 'Some guy published a prediction of future Nobels – I think it was in *Omni* magazine – and he listed us,' says Wilson. 'But he'd been wrong on a bunch of things, so Arno and I didn't take it seriously.' In the summer of 1978, there was another hint, this time from an Irishman who had worked at Bell Labs. Jerry Rickson, while visiting Sweden, had been buttonholed by one of the country's leading radio astronomers. 'He got asked some very detailed questions about Arno and me and our relationship,' says Wilson. 'Who did what – that sort of thing.'

Later, a Swiss colleague of Wilson's was more blatant. Martin Schneider was late handing Wilson a progress report on an experiment, and when the pair ran into each other at Bell Labs, Wilson asked whether he could have the report on his desk the next day. 'You won't want it tomorrow,' Schneider said, gleefully. 'They're going to announce your Nobel Prize!'

The next morning, the phone jangling woke him at 7am. It was another of his colleagues at Bell Labs; he had heard a news item on WCBS radio and wanted to know whether it was true what people were saying, that he and Arno Penzias had won the Nobel Prize? Wilson could not say for sure, but all doubt was removed with the arrival of a telegram from the Royal Swedish Academy of Sciences. The 1978 Nobel Prize in Physics had been awarded to Penzias and Wilson for their discovery of the three-degree cosmic background radiation.

For Alpher and Herman, the award of the Nobel Prize in Physics added yet more salt to the wound. Penzias and Wilson had stumbled on the radiation they had predicted seventeen years earlier entirely by accident. And if that was not bad enough, the Bell Labs researchers had not admitted for two years afterwards that the signal had anything whatsoever to do with the birth of the universe.

Dicke and his colleagues maintained they had been unaware of Alpher and Herman's 1948 prediction of the Big Bang afterglow, but they failed to put the record straight. To be fair, they tried on several occasions, but in Alpher and Herman's opinion they did not try hard enough.* So the wound remained. As for Gamow, he remained bitter about the treatment he and his team had suffered until his premature death from alcohol-induced liver disease in 1968. His only good fortune was not to be around to hear of the award of the 1978 Nobel Prize in Physics.

Wilson, of course, had no power over the Nobel Committee. 'I consider myself so lucky,' he says.

o   o   o

The afterglow of the Big Bang is the single most striking feature of our universe. If we had eyes that could see microwaves rather than visible light, we would see all of space glowing a dazzlingly brilliant white. It would be like being inside a giant light bulb. The question therefore arises: Why did it take until 1965 for the cosmic background radiation to be discovered by Penzias and Wilson – and then only by accident? The Nobel Prize-winning physicist Steven Weinberg has thought long and hard about

* Dicke believed in an oscillating universe, in which the cosmos repeatedly expanded and contracted throughout eternity like a giant beating heart. If each cycle were to begin like the previous one, it would be necessary to destroy the elements built up in the previous cycle, probably inside stars. Dicke realised that extreme heat would do the trick, by slamming together nuclei so violently that they disintegrated into hydrogen. Thus he hit on the idea of the universe going through a hot dense phase, with heat radiation left over, for exactly the opposite reason to Gamow. This is the way real science is done; both men were right for the wrong reasons, and each wrong reason was different from the other. After all, we do not appear to live in an oscillating universe (as Dicke thought) and most of the elements were not forged in a hot Big Bang (as Gamow thought).

this question and about why there was no earlier systematic search. In his popular account of the Big Bang, *The First Three Minutes*, he proposed three main reasons.

Firstly and most obviously, says Weinberg, Alpher and Herman were told by radio astronomers that the microwave afterglow of the Big Bang was undetectable. This was incorrect. Detecting it would admittedly have been hard, requiring a cold load with which to compare the temperature of the sky, but it could have been done.

The second reason why nobody looked for the fireball radiation, Weinberg says, is that its prediction emerged from a theory which was later discredited. By the 1950s, it was clear to everyone that most elements could not, as George Gamow had hoped, have been made in the Big Bang. Nature used two main furnaces to forge the elements: the fireball of the Big Bang, which made helium and the lightest elements in the first few minutes of the universe's existence; and the stars, which subsequently created all the heavier elements. Unfortunately, when it became obvious that the Big Bang could not have made nature's heavy elements, the idea was abandoned and Gamow's baby was thrown out with the bathwater.

But the most important reason why the Big Bang theory did not lead to a search for the fireball radiation, Weinberg says, was that before 1965 it was extraordinarily difficult for physicists to truly take seriously a theory of the early universe. 'The mistake of physicists is not in taking their theories too seriously but in not taking them seriously enough,' says Weinberg.[16] It was a simple failure of imagination. The temperature and density of matter in the first few minutes of creation were so far removed from everyday experience that it was hard for anyone to believe that they had ever actually occurred. Scientists could not imagine that anything as stark staring bonkers as the Big Bang could really be true. 'The most important thing accomplished by the ultimate discovery of the three-kelvin radiation background was to force all of us to take seriously the idea that there was an early universe,' says Weinberg.[17]

Vandenberg Air Force Base, California, 18 November 1989

The night before the launch of NASA's Cosmic Background Explorer satellite, the COBE team flew out to Vandenberg Air Force Base, one hundred miles north of Los Angeles. They were put onto buses at around 3am, which put them down in a field about a mile from the launch pad. It was freezing and dawn was still some time away.

COBE, with its on-board microwave horns, was designed to observe the cosmic background radiation from above the Earth's atmosphere, which made it difficult to observe from the ground. Impressed on the radiation is a 'baby photo' of the universe when it was just 380,000 years old and the matter of the cooling Big Bang was just beginning to clump together under gravity, to form what would eventually become galaxies. COBE was going to take that photo.

It was a large gathering and there was great excitement and anticipation. In the waiting crowd, stamping their feet to keep warm, were two elderly men who had been both surprised and pleased to be included. John Mather, project scientist and leader of the COBE team, had made a special point of inviting Ralph Alpher and Robert Herman. At long last, everyone recognised the prescience of the two magicians in predicting the afterglow of the Big Bang in 1948.

7

The holes in the sky

The black holes of nature are the most perfect
macroscopic objects there are in the universe:
the only elements in their construction are
our concepts of space and time.

SUBRAHMANYAN CHANDRASEKHAR

Black holes are where God divided by zero.

STEVEN WRIGHT

Herstmonceux, Sussex, Autumn 1971

There was something terribly wrong about the blue star. It seemed to be orbiting a body that was not there. Two astronomers sat at a desk in an octagonal turret room of a fifteenth-century English castle and surveyed their puzzling observations.

It was the autumn of 1971, and Paul Murdin and Louise Webster had been sharing an office at the Royal Greenwich Observatory (RGO) at Herstmonceux Castle ever since the summer, when Murdin had returned from a seven-year stint in America. It was a large room, entered through a wooden door so low that Webster had to duck. When the castle had been built, the room had possessed only vertical slits, but a larger window had been added later. Through it, the moat that surrounded the picturesque castle could be seen and, beyond that, pastures dotted with grazing geese. On the wall beside the window, previous occupants of the turret room had recorded the date each year when the swifts returned to the castle.

The observations the two astronomers were poring over were of a star called HDE 226868. HDE stood for 'Henry Draper Extension', a catalogue of stars compiled between 1925 and 1936 by a small army of female astronomers at Harvard University and named after and paid for by the widow of an American doctor and amateur astronomer. Murdin had first become aware of HDE 226868 while at the University of Rochester in New York state, where he had carried out his PhD research and held several short-term postdoctoral jobs.

The big problem in astronomy, Murdin had realised while a student, was identifying something worth studying. As British humorist Douglas Adams remarked, 'Space is big. You just won't believe how vastly, hugely, mind-bogglingly big it is. I mean, you may think it's a long way down the road to the chemist's, but that's just peanuts to space.'[1] There are about two trillion galaxies, and many of those contain hundreds of billions of stars. Finding an interesting object is tantamount to finding an interesting grain of sand among all the ordinary grains on all the beaches of the world.

What was needed, Murdin realised, was some observational signature that might flag the fact that something unusual was going on. Together with the other graduate students at Rochester, he had often debated what might constitute such a sign; the emission of radio waves was one possibility, but Murdin had a nose for an interesting idea and zeroed in on the emission of X-rays as the newest and most promising sign.

X-rays are a high-energy type of light given out by matter when it is heated to hundreds of thousands, or even millions, of degrees. The Earth's atmosphere screens out X-rays from space, which is unfortunate for astronomers, though not for life on Earth. However, in the late 1950s and early 1960s, the

Italian–American physicist Riccardo Giacconi built the first crude X-ray telescope, which he and his colleagues launched on 'sounding rockets' that climbed above most of the atmosphere, before falling back to Earth. Their brief glimpses of the X-ray universe revealed a number of cosmic sources, and among them was Cygnus X-1, discovered in 1962, one of the brightest objects in the sky.

Unfortunately, early X-ray telescopes were so crude that they could rarely pin down where a cosmic source was more precisely than 'in a particular constellation'. In the case of Cygnus X-1, the constellation was Cygnus, the swan. No one knew what a star emitting X-rays would look like, so the idea was to look for something unusual. At first, the area of the sky to be searched was so large that the task was all but hopeless, but by 1970, technical improvements in X-ray telescopes had improved the situation. Murdin noticed that the 'box' showing the possible location of Cygnus X-1 contained one star that was far brighter than the rest: HDE 226868. However, there did not seem to be anything strange about it.

By the time Murdin returned to England – to yet another short-term post, this time at the RGO at Herstmonceux in East Sussex – NASA had launched the first satellite carrying an astronomical telescope that was sensitive to X-rays. 'Uhuru' located a host of celestial X-ray sources, and Murdin obtained a preprint of its catalogue. The box pinning down Cygnus X-1's location had shrunk considerably and was now only about a third of the apparent diameter of the full Moon. And crucially, it still contained HDE 226868. 'The star was still waving a flag saying, "Look at me! I'm actually quite interesting!"' says Murdin.

There things might have stayed had it not been for a piece of good fortune: Murdin sharing an office with Webster. An

Australian just a tad older than Murdin, who was not yet thirty, Webster was working with Richard Woolley, director of the RGO, on a project to study how the stars move through space in our galaxy, the Milky Way.[2]

Murdin did not think the blue star HDE 226868 itself was the source of the mysterious X-rays – he and Webster had studied the star's light and it was too ordinary.[3] However, it was suspected that X-rays might be generated by matter from one star swirling down onto a compact, superdense companion, rather like water swirling down a plughole. Internal friction in such an 'accretion disc' would make the matter so hot that it glowed with X-rays. The question in Murdin's mind was therefore: Did HDE 226868 have a companion that might be the source of the X-rays?

If the blue star varied its speed over time – approaching the Earth before later receding from it – it would be a sure sign that it was circling another star. Murdin did not even have to do any work to find out whether the star's speed was changing, since his office mate and her team were already measuring the speeds of stars; he simply wrote the celestial co-ordinates of HDE 226868 on a filing card and handed it to Webster.

Webster was using the giant hundred-inch Isaac Newton Telescope, which was ridiculously situated at Herstmonceux, a location plagued by cloud, mist and rain (in 1979, at a cost greater than that required to construct it in the first place, it would be moved to the Roque de los Muchachos Observatory on La Palma in the Canary Islands). But this, it turned out, was another piece of good fortune for Murdin. Whereas the world's best observatories, which were often at high-altitude, dry sites, where the 'seeing' was excellent, concentrated on the faintest and most distant objects in the universe, at Herstmonceux there was no choice but to focus on brighter objects, which could be picked out even in a murky sky. HDE 226868 was a good candidate.

Webster was using a super-sensitive 'spectrometer' to obtain the spectra of the stars. A star's spectrum is a record of how its brightness varies with the frequency, or colour, of its light. The frequency is like the pitch of a sound; it gets higher when a star is moving towards us and lower when it is moving away from us. This phenomenon, a direct analogue of a police siren sounding shriller as it approaches and deeper as recedes, is known as the 'Doppler effect'.

Fortunately, nature has given each type of atom a set of characteristic frequencies at which it gives out light, which acts like a fingerprint. To detect the motion of a star along the line of sight, it is necessary only to identify one such spectral feature in the star's spectrum and then see whether it is shifted from the frequency it would have in a laboratory back on Earth.

Webster and her team took six spectra of HDE 226868; disappointingly, only one of the six showed any sign of motion. It was not promising, and Murdin was beginning to lose interest in the blue star. But since he was not the person doing the legwork, and there was a lone spectrum that offered some sort of promise, he decided to leave the filing card where it was and the blue star stayed in Webster's observing programme.

Webster duly took another set of spectra, but when they came back, Murdin was astonished to see that they revealed motion. It was immediately obvious that HDE 226868 was in orbit around another star. When Murdin did the calculations, cranking the handle of what would today be regarded as a prehistoric mechanical calculating machine, he found it was circling once every 5.6 days.

The orbit deduced by Murdin showed why Webster's first spectra had shown so little sign of motion. The Doppler shift reveals only the component of a star's motion towards and away from the Earth. But by a piece of bad luck, five of the first set of spectra had been taken at a time in the orbit of HDE 226868 when it was travelling across the line of sight and so barely moving towards or away from us.

With hindsight, Murdin could see that the sole spectrum in the first set that provided a positive result was an erroneous one, caused by some instrumental problem that would never be tracked down. But it was another piece of good luck for Murdin; if it had not been for the sixth spectrum and its anomalous indication of motion, he would not have left the filing card with Webster and would not have discovered the companion of HDE 226868.

As Murdin and Webster sat with the latest data on their desk, their focus was on the mass of the unseen companion star. Having now obtained multiple spectra of HDE 226868, the two astronomers knew it was an extremely young, extremely hot and extremely luminous star, pumping out about 400,000 times as much light as the Sun. In 1971, the average mass of such a 'type-O blue supergiant' was believed to be about twenty solar masses. Using this mass and the orbital period of 5.6 days, it was possible to determine the mass of the companion. Actually, that was not entirely true – because the orbit of HDE 226868 was projected onto the two-dimensional sky and its true orientation in three-dimensional space was unknown, it was possible to say only that the companion was bigger than a certain 'minimum mass'.

Murdin and Webster divided up the calculating work between them. Murdin went off to consult reference books in the library fifty metres down the corridor, double- and triple-checking that he was using the correct formula for deducing the unknown mass. It was a complicated expression and he wanted to make sure he had remembered it correctly.

The meeting between the two astronomers had lasted only an hour, but they were certain of their conclusion. The companion was at least four times as massive as the Sun and probably six times as big.[4]

At the time, the only compact stellar objects that were known were 'white dwarfs' and 'neutron stars', the latter having been discovered four years earlier in the guise of 'pulsars' by Cambridge graduate student Jocelyn Bell. But theory constrained both types of body to weigh less than a couple of solar masses, which left only one type of theoretical object as a candidate for the invisible star. Webster and Murdin looked at each other. She was calm and unflappable, as always, while he could hardly contain his excitement. The object they were both thinking of was a monstrous nightmare entity, whose existence had been predicted more than half a century earlier by a man dying in a field hospital bed . . .

The Eastern Front, Winter 1916

The sound of the guns woke him. Karl Schwarzschild could feel the dull pounding deep in his bones as he screwed his eyes closed against the winter sunshine streaming through the thin curtains. For a moment his mood was black. With the pain and discomfort he had been suffering, it had been hard to get to sleep, but he could not permit himself to succumb to self-pity; that way lay oblivion. He must cling at all costs to the good things. He glanced across to where his calculations lay on his bedside cabinet, terrified for a split second that they had been nothing but a dream. But no – thank goodness – the scientific paper he had written was exactly where he had left it in the early hours of the morning. Everything was true. With a pen and paper, he had revealed something extraordinary about the universe: that somewhere out in space there might exist celestial bodies so monstrous they were the stuff of nightmares.

The morning routine was exhausting and time-consuming. Nurses in starched white uniforms entered his isolation room. Kindly and gently, they mopped at the ugly weeping blisters that now covered most of his body and sat him in a chair while they changed his bloody sheets, before leaving him with a tray of soft bread and warm milk (though he would have preferred a beer).[5] As he chewed at the soggy bread – the only foodstuff that did not further inflame his blistered mouth – he listened to the thud of the guns and pondered the chain of events that had brought him to this hospital on the Eastern Front.

When war was declared in August 1914, there had been no need for Schwarzschild to volunteer. Not only was he a man of forty but, as director of the Berlin Observatory, he held one of the most prestigious jobs in German science. But anti-Semitism was on the rise, and he was a Jew. It was not something he advertised. In fact, in his will, which he had penned on the eve of joining up, he strongly advised his wife not to tell his children he was Jewish until they were at least fourteen or fifteen.[6] But though he lived a secular life and did not attend synagogue, he nevertheless felt very strongly that it was necessary to stand up and be counted; if Jews were to push back the tide of anti-Semitism, they must demonstrate beyond any doubt that they were patriotic Germans. That was why, as ominous events unfolded across Europe during the summer of

1914, he had resolved that if it came to it, he would put his life on the line to defend the Fatherland. In his eighteen months in the Kaiser's army, Schwarzschild had run a weather station in Belgium, calculated shell trajectories with an artillery battery in France and, finally, been posted to the Eastern Front. It was there he developed the mouth ulcers. At first he thought they were the result of exhaustion. The winter of 1915 had been cold, and it was stressful being apart from his wife and family. But the ulcers had worsened, and within a month, blood-filled blisters had erupted all over his body, forming large areas of painful, raw-looking sores that eventually formed scabs. The blisters seemed to come and go in waves, flaring up and dying down unpredictably.

When he had arrived at the field hospital, the doctors were completely puzzled. It was several days before they diagnosed *pemphigus vulgaris*. Nobody knew what triggered the rare condition in which the body attacks its own skin, but it was known to be more common among Jews, particularly Ashkenazis from Eastern Europe.[7] He was told that there was no known cure.

As a scientist, Schwarzschild wanted to know everything. The doctors, perhaps to spare him, were evasive, but it was patently obvious to him that his condition was potentially life-threatening. The skin, after all, is the body's largest organ. It is through the skin that we sweat, and if it is compromised, the body has no means of avoiding over-heating. Not only that but the skin provides a barrier against infection; if it is breached, the body is left open to attack by every alien microorganism.

The paper he had just finished was the second he had begun while with his artillery battery at the Eastern Front. He began to check through the calculations. Had he made a mistake? Did his results hold up? He had no one to share his thoughts with.

These past weeks he had known what it felt like for Newton, 'voyaging through strange seas of thought, alone'.[8] So fresh was the new theory of gravity he was using that he knew he was one of the first people, if not the very first, to understand and master it. Apart, of course, from its genius creator.

o   o   o

Albert Einstein had crossed paths with Schwarzschild on only a handful of occasions, and they had exchanged little more than pleasantries. The reason was that the Berlin Observatory was in Potsdam, just outside the city, while Einstein's workplace, the Kaiser Wilhelm Institute for Physics, was in the suburb of Dahlem, nearer the centre. Despite this minimal contact, however, Schwarzschild had followed Einstein's decade-long struggle to find a theory of gravity that was compatible with his special theory of relativity with 'burning interest'.[9]

The special theory of relativity contradicts Newton's theory of gravity in several ways. Whereas Einstein recognised that nothing can travel faster than light – the cosmic speed limit – Newton assumed that the gravity of a body like the Sun is felt everywhere instantaneously, which is equivalent to saying that gravitational influence propagates at infinite speed. And whereas Einstein recognised that all forms of energy are sources of gravity because all forms of energy have an effective mass, Newton assumed that mass alone is a source of gravity.*

The fact that light energy has an effective mass has an observable consequence, which Einstein realised before he achieved his goal of finding a theory of gravity that was compatible with

---

* Or, strictly speaking, mass-energy.

relativity. The path of starlight that passes close to the Sun on its journey to the Earth should be bent by solar gravity, and when the First World War broke out in 1914, Schwarzschild's colleague Erwin Freundlich had been in the Crimea with two companions, planning to observe light bending during the total solar eclipse of 21 August.[10] Unfortunately, they were thrown into prison by the Russians as enemy aliens and had limped back to Berlin in late September, in one of the first prisoner exchanges of the war.

Einstein's struggle to find the elusive theory of gravity culminated in his presentation of the 'general theory of relativity' in four papers to the Prussian Academy of Sciences in November 1915. It painted a picture of a new and unexpected world.

According to Newton, a 'force' of gravity exists between the Sun and the Earth, much like an invisible tether that connects the two bodies and keeps the Earth perpetually trapped in orbit around the Sun. Einstein showed this is wrong; what a mass like the Sun actually does is create a valley in the space–time around it.[11] The Earth then traverses the upper slopes of this valley like a ball in a roulette wheel. Though no one would use such words for another half a century, what the general theory of relativity says in essence is: 'Matter tells space–time how to warp, and warped space–time tells matter how to move.'[12]

Warped space–time, according to Einstein, *is* gravity. However, since space–time is a four-dimensional thing and we are three-dimensional beings, we are utterly unaware of the hills and valleys of space–time. To explain the motion of a body like the Earth around the Sun, we have, therefore, invented a 'force' called gravity.

Copies of Einstein's papers on his new theory reached Schwarzschild on the Eastern Front within days of their presentation in Berlin. He had instantly fallen in love with the general theory of relativity.[13] Its beauty and daring took his breath away. But more importantly, it took him to another place, far away from the death, destruction and the pounding of the guns. Incredibly, he found time between his calculations of artillery trajectories to absorb the complex mathematics and think deeply about the consequences of the theory.

Einstein had used his theory to explain the baffling motion of the planet closest to the Sun. Mercury, like all the planets, is tugged by the gravity not only of the Sun but also of the other planets in the solar system. They cause its elliptical orbit to gradually change its orientation in space, or 'precess'. But even when this effect is taken into account, there remains an unexplained bit left over: the 'anomalous precession of the perihelion of Mercury'.

Einstein realised that Mercury's proximity to the biggest mass in the solar system meant that it was orbiting in the most warped space–time of any planet and that its path through space would therefore be affected by this. He used his theory of gravity to predict the path and found that it was exactly what was observed by astronomers. It was a triumph. His theory perfectly explained the anomalous motion of Mercury.

However, Einstein's calculations were scrappy and inelegant. The problem was that the machinery of his theory of gravity was complex. It used the mathematics of curved space–time, which had been developed in the nineteenth century by a number of mathematicians, most notably Carl Friedrich Gauss and Bernhard Riemann. Whereas one equation is sufficient to describe gravity in Newton's theory, Einstein's requires a total of ten.[14] Consequently, finding the shape of the space–time for a given distribution of matter – known technically as a 'solution' of Einstein's equations of the gravitational 'field' – is hard. Einstein himself had believed it impossible, so, in explaining the anomalous motion of Mercury, he had resorted to an approximate expression for the curved space–time around the Sun.

Schwarzschild was familiar with 'Riemannian geometry': the mathematics of curved space. While the guns boomed around him, he wondered whether he could do better than Einstein. Could he find an exact formula for the curvature of space around a localised mass like the Sun?

He started by making some basic assumptions. First, that the Sun – or indeed any star – is perfectly spherical. Secondly, that the curvature of the space–time around it does not change with time. And thirdly, that the curvature of space–time does not depend on direction, but only on the 'radial' distance from the Sun. Remarkably, these insights allowed Schwarzschild to hugely simplify Einstein's equations, reducing them from ten to just one. He then employed a little mathematical wizardry, and – miracle of miracles – discovered that the lone equation had a unique solution.

Schwarzschild had achieved the impossible: he had out-Einsteined Einstein. Rather than an approximate expression for the curvature of space–time surrounding the Sun, he had found a precise description, which was the first exact solution of Einstein's theory of gravity ever found. In years to come, in recognition of the difficulty of finding a solution to Einstein's equations, physicists would refer to such solutions by the names of their discoverers. His would be immortalised as the Schwarzschild solution, or, more precisely, the 'Schwarzschild metric'.

Using his exact solution, Schwarzschild quickly confirmed Einstein's claim that his theory explained the anomalous motion of Mercury. 'It is quite a wonderful thing that from such an abstract idea the explanation of the Mercury anomaly emerges so inevitably,' he observed.[15]

Schwarzschild then wrote up his calculations as a paper, and on 22 December 1915, wrote a covering letter to Einstein. It concluded, 'As you see, the war treated me kindly enough, in spite of the heavy gunfire, to allow me to get away from it all and take this walk in the land of your ideas.'[16] As yet, he had not realised the seriousness of the blisters which had begun to form in his mouth and would soon see him invalided out of the army to a field hospital.

o   o   o

It was a surprise to receive a letter from the Eastern Front. Einstein was aware – because it was such an extraordinary thing – that the director of the Berlin Observatory, despite being forty years old, had volunteered for the Kaiser's army at the outbreak of war, but what could he be writing to him about?

On reading, he was astonished to find a calculation using his own theory. He ran his finger along the lines of algebra, nodding emphatically as he did so. He had presented his theory of gravity to the Prussian Academy of Sciences little more than a month ago, yet Schwarzschild had not only mastered it but driven it forward into new territory. Here was the first exact solution of his general theory of relativity, something Einstein himself had considered impossible.

Immediately, he replied to Schwarzschild, 'I have read your paper with the utmost interest. I had not expected that one could formulate the solution to the problem in such a simple way. I liked very much your mathematical treatment of the subject.'[17]

Einstein promised to present the work to the Prussian Academy on the following Thursday, along with a few words of explanation. He was as good as his word and delivered a summary of Schwarzschild's paper on 13 January 1916. But Schwarzschild, lying in a hospital bed, had not finished with Einstein's theory. He had examined the case of an idealised star – a spherical mass – and found an exact description of the curvature of space–time on the outside, but what about the inside? That was the subject of his second paper, whose calculations he was now checking, and which he was on the verge of sending to Einstein.

The topic had captivated him for several days, and most importantly it had taken away his pain. He was oblivious to everything, a man lost in the dream. 'Professor Schwarzschild!' he remembered them shaking him. 'We need to change your dressing . . . your bedclothes . . . You need to go for a walk . . .'

What he had discovered by perusing his miraculous solution was something incredible. If a celestial body were ever to be compressed within a certain critical radius, space–time would become so grossly warped that it would no longer be a mere valley.[18] Instead, it would morph into a bottomless pit from which nothing, not even a beam of light, could ever climb out. The star would become cut off from the universe forever and would appear like a hole in space. He had no name for such a region of grossly warped space–time, but one day there would be almost nobody on Earth who had not heard of the term 'black hole'.

The threshold radius was ridiculously small. Just like Schwarzschild's space–time solution itself, it would one day bear the name of its discoverer. For the Sun, the 'Schwarzschild radius' was 1.47 kilometres, and for the Earth it was a mere five millimetres. If the Sun and the Earth were squeezed this small, they would wink out of existence, disappearing from view forever.

But the Sun is more than a million kilometres across, so this would mean compressing its material to an enormous, mind-boggling density. Schwarzschild's first reaction was that this was 'very weird and perhaps just a mathematical curiosity', but he did not dismiss it out of hand.[19] 'History tells us that the mathematical solutions are often realised in nature, as if there were some kind of pre-established harmony between mathematics and physics,' he wrote. This idea had been quite strongly present at the University of Göttingen, where he had worked before Berlin, and he confessed to being a 'believer'. Maybe the monster described by his equation might actually exist.

Schwarzschild folded the letter containing his new paper, slipped it in an envelope and sealed it. When an orderly came by, he gave it to him to post.

o   o   o

Einstein presented Schwarzschild's black hole solution to the Prussian Academy of Sciences on 13 February 1916. In March, Schwarzschild, whose condition had worsened, was moved to Berlin, and he died on 11 May 1916. He was just forty-two. But one thing did not die with him: his solution of Einstein's equations for a black hole.

Herstmonceux, Sussex, Winter 1971

After the meeting with Louise Webster, Paul Murdin found it impossible to settle: it was the adrenalin coursing through him. Webster, as reserved and unperturbable as ever, worked at her desk, immune to distraction, while Murdin paced up and down in the turret room, going over the logic of the remarkable conclusion they had come to.

The X-rays in Cygnus X-1 came from matter torn off the blue supergiant star and heated to incandescence by internal friction as it swirled down into a black hole. If the two astronomers were right, they had made a truly momentous astronomical discovery. Nevertheless, it was still hard to believe that such a wild theoretical prediction could come true. 'The surprising thing is that black holes turn out to be real objects,' remarked Murdin. 'Incredibly, they actually exist!'

Murdin hoped that finding the first black hole out in space would make his name in astronomy and that, even more importantly, it might lead to a permanent job. With a young family to support, that was more than a trivial consideration.

Murdin and Webster wrote a 500-word paper on their discovery, but they had a surprising amount of difficulty in getting permission to send it to *Nature* for publication. The Royal Observatory's director, Richard Woolley, did not believe in black holes, thinking they were some sort of 'new-age-ish' magic. 'In one conversation he even asked me why, exactly, I believed that Cygnus X-1 contained a "black box",' says Murdin.

Part of the reason for Woolley's foot-dragging was that he had been the student of Arthur Eddington, who had not believed in black holes, but another reason was that the Royal Greenwich Observatory at Herstmonceux had until recently been run by the Royal Navy. Public perception was extremely important to the navy, and Woolley was fearful that the Observatory might make a claim that would open it up to derision and ridicule.

But the observations Murdin and Webster had accumulated were nothing if not convincing: all the evidence pointed to

HDE 226868 being in orbit around something invisible and massive, and the only conceivable object that fit the bill was a black hole. Finally, after consulting with other senior members of the RGO, Woolley caved in and gave permission to Murdin and Webster to submit their paper for publication.

A lot was at stake for Murdin and it was a nail-biting time. There was no guarantee that someone else would not come to the same conclusion about HDE 226868 and beat him and Webster into print. To guard against this, Murdin decided to deliver the paper personally to the *Nature* offices in central London, making sure it was stamped with the date. However, while driving to Hastings Station to catch the train, he half heard a news item on his car radio that seemed to be about a very energetic event in the stars. He immediately thought, 'Oh no, someone else has got our fantastic result! We've been scooped!'

All day in London, Murdin was worried sick. It was only when he returned to Hastings that evening that he heard the news item repeated; to his enormous relief it turned out to be about a storm on Mars.

The paper appeared in *Nature* on 7 January 1972.[20] Murdin got his permanent job and his family got a bigger house. In fact, he was the first person in history to have his mortgage paid by a black hole.[21]

The day the paper was published, Murdin and his wife Lesley celebrated by taking their two small boys to a café on Hastings seafront and treating them to 'knickerbocker glories'. As the boys, aged three and seven, dug their long spoons into the layers of ice cream, fruit and syrup, it was not difficult to guess what was going through their minds. 'I think they were hoping their dad would find some more black holes,' says Murdin.[22]

It would be hard to imagine a greater contrast between the world of the man who had co-discovered black holes and that of the magician who had predicted the existence of black holes from his hospital bed on the Eastern Front fifty-six years earlier.

o   o   o

By the time Murdin and Webster discovered the first black hole in Cygnus X-1, the theory of such objects had moved on from Schwarzschild's exact solution to Einstein's theory of gravity.[23] Einstein never believed in the possibility of black holes, and most others who considered the solution shared his view – though not Schwarzschild – that a body from which nothing, not even light, can escape is simply too weird for words. When a massive star shrinks at the end of its life, they reasoned, some as-yet-unknown force must intervene to prevent the formation of such a monstrosity. Such a force appeared to be provided by a revolutionary new description of the world of atoms and their constituents, which was formulated in the 1920s.

'Quantum theory' recognises that the fundamental building blocks of the world such as atoms, electrons and photons have a weird 'dual' nature.[24] They can behave both as particles – like tiny billiard balls – and waves, like ripples on a pond. Because such quantum waves are spread out and so need a lot of room, the particles with which they are associated are hard to squeeze into a small volume. Or, to put it another way, when they are compressed, they push back.

It turns out that the smaller the particle, the bigger the quantum wave. The smallest familiar particle, with the biggest quantum wave, is the electron, so when the matter of a star is squeezed into a small volume, the electrons that orbit inside its atoms push back. This electron 'degeneracy pressure' is the force that intervenes to stop a star shrinking to form a black hole. Or so people thought.

In 1930, a nineteen-year-old physics student, travelling by sea from India to England, showed that everything is changed by Einstein's special theory of relativity. Subrahmanyan Chandrasekhar imagined things from the particle viewpoint. Electrons push back when squeezed into a small volume because they buzz about ever faster, like a swarm of angry bees. However, Einstein's theory recognises that nothing can travel faster than light, so there is a limit to how fast electrons can go and how hard they can push back. If a dying star is less than about 1.4 times the mass of the Sun, electron degeneracy pressure can indeed hold gravity at bay, resulting in a highly compact 'white dwarf', but for a star above this 'Chandrasekhar limit', things are different. Its gravity is strong enough to overcome the buzzing of its electrons, so nothing can prevent its runaway shrinkage to form a black hole.

A twist to the story was added with the discovery of the neutron by James Chadwick in 1932. Together with protons, neutrons compose an atom's central 'nucleus', around which electrons whirl like planets around the Sun. If a star is squeezed into a small enough volume, its electrons are squeezed into its protons, creating a dense ball of neutrons. The neutrons of such a 'neutron star', like electrons, have a quantum wave associated with them and resist being squeezed. But like electrons, there is a limit to how fast they can buzz about and how hard they can push back. The effect is more complicated to calculate than for electrons because it involves the 'strong nuclear force', which is hard for theorists to model. But for a star above about three times the mass of the Sun, gravity is strong enough to overcome the buzzing of neutrons and nothing can stop the formation of a black hole. Black holes are unavoidable. And so too are the problems they create for physics.

The main reason black holes have been considered such monstrosities is that, when a star undergoes runaway shrinkage to form a black hole, it eventually ends up squeezed into an infinitesimal point, with its density sky-rocketing to infinity. Such a 'singularity' signals the breakdown of space and time – indeed of physics itself.

'Black holes are very exotic objects,' says Andrea Ghez of the University of California at Los Angeles. 'Technically, a black hole puts a huge amount of mass inside of zero volume. So our understanding of the centre of black holes doesn't make sense, which is a big clue to physicists that we don't have our physics quite right.'[25]

The American physicist John Wheeler put it more poetically: 'The black hole teaches us that space can be crumpled like a piece of paper into an infinitesimal dot, that time can be extinguished like a blown-out flame, and that the laws of physics that we regard as "sacred", as immutable, are anything but.'[26]

No wonder Einstein abhorred the fact that his theory of gravity predicted the existence of such monsters – it contains within it the seeds of its own destruction. To understand what really happens to space and time at the heart of a black hole, it will be necessary to find a deeper, singularity-free theory of gravity. Einstein's theory of gravity is expected to be an approximation of this deeper theory, just as Newton's theory of gravity turned out to be an approximation of Einstein's.

A black hole's singularity is, however, surrounded by a 'horizon', an imaginary membrane that marks the point of no return for in-falling light and matter. But it is no mere passive boundary, and in 1974, British physicist Stephen Hawking discovered something remarkable about it.

To appreciate what Hawking discovered, it is necessary to understand what quantum theory says about empty space. Far from being empty, it is seething with energy; subatomic particles and their antiparticles are continually popping into existence in pairs, as permitted by the 'Heisenberg Uncertainty Principle'. Nature turns a blind eye to these particles, not bothering about where the energy to create them comes from, as long as they meet and destroy, or 'annihilate', each other very quickly. It is a bit like a teenager borrowing a car from their dad overnight and returning it to the garage before he notices it is missing.

But Hawking realised that, near the horizon of a black hole, something very interesting happens. There is the possibility that one of the particles of a newly created pair falls through the horizon into the black hole. The remaining particle has no partner with which to annihilate and flies away from the hole, along with countless others in the same situation. Contrary to all expectations, therefore, black holes are not black but glow with emitted particles – 'Hawking radiation'.

Hawking had earlier discovered that when black holes merge, the surface area of the horizon of the merged hole is always bigger than the sum of the areas of the two precursor black holes. The Israeli physicist Jacob Bekenstein had speculated that the surface area represents the 'entropy' of the black hole. This is a property that arises in the theory of thermodynamics – the theory of heat and motion that underpins physics, chemistry and many other fields – and which always increases. But it applies only to hot bodies, so how could it possibly apply to a black hole?

Hawking had found the answer: thermodynamics applied to black holes because they are hot. The proof is that they glow with heat – Hawking radiation. The significance of Hawking's discovery is that at the horizon of a black hole, three of the great theories of physics meet – Einstein's theory of gravity, quantum theory and thermodynamics. It was a first tentative step on the road to uniting them – the holy grail of physics.

However, Hawking radiation threw up a serious problem: its particles do not come from inside a black hole since nothing can escape its gravity. Instead, they are created just outside the horizon. The energy to create them has to come from somewhere, and it comes from the gravitational energy of the black hole itself. As it radiates Hawking radiation, it gradually shrinks away. Star-sized black holes have extremely weak Hawking radiation but, as a black hole gets smaller, the radiation gets brighter, until the hole finally explodes in a blinding flash. Since such 'evaporation' would take far longer than the current age of the universe, it might seem of no consequence, but nothing could be further from the truth.

It is a cornerstone of physics that information cannot be destroyed. A complete description of the star that initially collapsed to form a black hole would require recording the type and position of each of the huge numbers of subatomic particles that compose it. But once a hole has evaporated, there is literally nothing left, so where does the information go?

The current speculation is that the horizon of a black hole is not smooth and featureless, as Einstein's theory of gravity suggests, but rough and irregular on the microscopic scale, and that it is in the lumps and bumps of its Lilliputian landscape that is stored the information that describes the star that gave birth to the black hole. Since Hawking radiation is born in the vacuum a hair's breadth above a black hole's event horizon, it stands to reason that it is influenced by the microscopic undulations of that membrane. Those undulations 'modulate' it in much the same way that pop music modulates the 'carrier wave' of a radio station. In this way, the information that described the precursor star is carried out into the universe, imprinted indelibly on the Hawking radiation. No information is lost, and one of the most precious laws of physics is left intact.

Since Murdin and Webster's 1971 discovery of the first black hole in Cygnus X-1, more candidates have been found, though the total stands at fewer than twenty-five. But black holes – of a very different kind – had been stumbled on almost a decade earlier, in 1963.

'Quasars', discovered by Dutch–American astronomer Maarten Schmidt, are the super-bright cores of newborn galaxies. They typically pump out one hundred times the energy of a galaxy of stars, but from a volume smaller than the solar system. The only possible source of such prodigious luminosity is matter, heated to incandescence, as it swirls like water down a plughole into a black hole. But not a stellar-mass hole – one with a mass up to fifty billion times that of the Sun.

Initially, it was thought that 'supermassive' black holes power only 'active galaxies', the 1 per cent of unruly galaxies, of which quasars are the most striking example. But in the 1990s, astronomers using NASA's Hubble Space Telescope in Earth's orbit discovered that there is a supermassive black hole lurking at the heart of pretty much every galaxy. The one in the core of the Milky Way, known as Sagittarius A*, is a tiddler, weighing in at only 4.3 million times the mass of the Sun. Why there should be a supermassive black hole in every galaxy remains one of the great unsolved mysteries of cosmology.

But despite the observational evidence for the existence of black holes, it was always indirect. Astronomers observed stars or hot gas whirling fantastically quickly around an unseen compact object and inferred the existence of a black hole, but there was always the possibility it was instead some undreamt-of super-compact object, held up by some hitherto unknown force.

The definitive proof of the existence of black holes came, however, on 14 September 2015. That was the day that gravitational waves – ripples in the very fabric of space–time, predicted by Einstein in 1916 – were detected on Earth for the first time. The key thing was that the waveform detected was precisely what Einstein's theory of gravity predicted for the merger of two black holes.

Black holes exist, beyond any doubt. Meanwhile, the quest continues to image them in space. The problem astronomers face is that stellar-mass black holes in the Milky Way are small and, well, black. Supermassive black holes, though big, are at cosmic distances and so also appear small. However, there are two black holes that are both relatively big and relatively nearby. One of them, Sagittarius A*, is 26,000 light years away in the centre of our galaxy, and the other, which is about one thousand times bigger, is in a nearby galaxy called M87.

Over the past few years, astronomers have been attempting to image the event horizons of these two supermassive black holes, using an array of co-operating radio telescopes scattered around the globe known as the 'Event Horizon Telescope' (EHT). The radio signals recorded at each site are combined on a computer at Haystack Observatory in Massachusetts, to simulate a giant dish the size of the Earth. The bigger the dish and the shorter the observing wavelength – EHT uses a wavelength of 1.3 millimetres – the more it can zoom in on details in the sky.

The hope is that the EHT will test a controversial recent claim by Hawking. Having shocked the world of physics in 1974 by claiming that black holes are not black but emit Hawking radiation, he did it again in 2014, when he claimed that event horizons do not exist, which means that, strictly speaking, neither do black holes.

The collapse of an object such as a star to form a black hole is violently chaotic and, rather than a horizon, all that forms, claimed Hawking, is a boundary of extreme space–time turbulence. Information can leak out through such an 'apparent horizon'. Hawking's conclusion was dramatic: 'The absence of event horizons means that there are no black holes – in the sense of regimes from which light can't escape to infinity,' he wrote. 'There are, however, apparent horizons which persist for a period of time.' Black holes, in other words, are not what we thought they were.

So is the horizon around a black hole the point of no return that everyone thought it was? Or is it merely an apparent horizon, as Hawking maintained, leaking stuff from inside the hole? The key thing is to observe the horizon and see whether it behaves as predicted by Einstein, or even whether it exists at all. 'An image will allow us to test general relativity at the black hole boundary, where it has never been tested before,' said Shep Doeleman of the Massachusetts Institute of Technology, leader of the EHT team. 'It would symbolise a turning point in our understanding of black holes and gravity.'

That turning point has now been reached. On 10 April 2019, the EHT team revealed the first-ever picture of a black hole.[27] It was not of Sagittarius A* but of M87, weighing in at seven billion times the mass of the Sun (Sagittarius A*, because it is smaller, was circled by matter many times while being observed, yielding a blurrier picture). The event horizon shows up as a dark 'shadow', backlit by intense radio waves emitted by matter heated to incandescence, as it swirls down through an 'accretion disc' onto the black hole.

'The hole is a part of our universe permanently screened from view,' says EHT physicist Feryal Özel of the University of Arizona in Tucson. 'A place where our physics – at least, as currently formulated – cannot reach.' Her Dutch colleague, Heino Falcke of Radboud University in Nijmegen, puts it more dramatically: 'We have seen the gates of hell at the very end of space and time.'

8

The god of small things

The discovery of the Higgs boson, just like the planet
Neptune and the radio wave, was first predicted with a
pencil, using mathematical equations.

MAX TEGMARK[1]

Tyger! Tyger! burning bright
In the forests of the night,
What immortal hand or eye,
Could frame thy fearful symmetry?

WILLIAM BLAKE[2]

Methodist Central Hall, London, 4 July 2012

Jon Butterworth was pissed off. He was pissed off because on
this day, which was destined to be one of the most remarkable
of his forty-five years, he was stuck in London. He did not
want to be in London. He wanted to be in Switzerland, where
the action was. And, as if to rub salt into the wound, he could
see the place where he would much rather be, on a giant video
screen at the back of the stage of Methodist Central Hall in
Westminster, London.

Butterworth's mood improved, however, as he sat down
beside his fellow panellists – his physicist colleague Jim
Virdee and John Womersley, CEO of the UK's Science and
Technology Facilities Council – and surveyed the audience.
Several hundred journalists, physicists and politicians, includ-
ing the science minister David Willetts, had crowded into this

one-hundred-year-old meeting place, a stone's throw from the Houses of Parliament, and the excitement and anticipation were palpable. Butterworth was pleasantly surprised that the public and the politicians alike seemed fascinated by what he and thousands of other physicists had been doing over much of the past decade.

Butterworth and Virdee were at this press conference in London rather than in Switzerland because they were the respective UK heads of ATLAS and CMS. These giant experiments at CERN, the European laboratory for particle physics near Geneva, were two of the 'eyes' of the Large Hadron Collider. They were located where its super-high-energy beams of counter-rotating protons slammed into each other. Each experiment consisted of detectors wrapped like layers of an onion around the collision points. They measured the energy, electric charge and direction of the myriad pieces of subatomic shrapnel that exploded outwards.

ATLAS had about three thousand physicists from thirty-eight countries working on it, and CMS had some 4,300 from forty-one nations. CMS was so massive that it weighed as much as the Eiffel Tower. Together, the ATLAS and CMS teams were looking for two distinctly different collision events, and each was the 'smoking gun' for the existence of a hypothetical subatomic particle. One like nothing ever seen before. One that was the key to understanding why the universe looks the way it does. And one whose existence had been predicted in a single sentence almost four decades before . . .

Edinburgh, August 1964

Peter Higgs was annoyed. His second paper in three weeks – on how nature's force-carrying particles might have acquired their masses – had been rejected by the editor of *Physics Letters*. The only thing that tempered his annoyance, as he looked out from his office at the city sparkling in the August sunshine, was his enormous relief that he was back here in his beloved Edinburgh, and no longer enduring the abject misery of the Western Highlands.

The camping trip had been a week ago but it still loomed large in his mind, like a nightmare that persists into the daylight hours. A friend had mentioned a place which, she had read in an article, had the lowest rainfall in Scotland, and Higgs had duly headed there with his American wife of one year Jody, whom he had met through their involvement with the Campaign for Nuclear Disarmament. Unfortunately, when they arrived, it was pouring with rain and they broke their borrowed tent while attempting to put it up. Sopping wet and bedraggled, they had holed up in a nearby bed and breakfast. To make matters worse, on their premature return to Edinburgh, their friend admitted that she had misread the article; far from having the lowest rain-fall in Scotland, the spot she had recommended had the highest.[3]

The camping trip had assumed mythic status in the minds of Higgs and his wife as one of those 'worst holiday ever' anec-dotes, guaranteed to elicit laughter when trotted out over drinks with friends. But for Higgs, the appalling weather was only part of the reason why the trip was so awful. Leaving his work half finished was the last thing he needed when he was on the verge of cracking the problem that had been his waking obsession for several years.

In his first paper, a mere one thousand words long, which he had completed on 24 July and submitted to the journal *Physics Letters*, he had shown how a promising theory of nature's fundamental forces that everyone believed was fatally flawed was not in fact flawed at all.[4] He had written it on a Monday after a light-bulb moment over the weekend and he considered it his very first real idea; in fact, the only idea he had ever had (he was a modest man).[5]

The paper finished on a cliff-hanger, with a tantalising promise of a second instalment to follow, but then had come the trip to the Highlands. The fact that his mind had so clearly been elsewhere had not helped his relationship with his wife. However, back in Edinburgh, warm, dry and free from distraction, he was able to apply total concentration to his work. He had quickly written, and on 31 July, submitted the second paper to *Physics Letters*.

When it had come back rejected, it had stung. There was no point pretending it had not. In his covering letter, Jacques Prentki, the journal's editor, suggested that Higgs do some more work on his theory. But the question was, what?

o   o   o

Higgs' two papers had had a long gestation. He was fascinated by quantum field theory, which melded into a coherent whole the two towering achievements of twentieth-century physics – Einstein's special theory of relativity, which describes what happens to space and time for bodies travelling close to the speed of light, and quantum theory, which describes the sub-microscopic world of atoms and their constituents. The physicist who had taken the first step towards such a unification was Paul Dirac, whose illustrious name appeared repeatedly on the board of honoured former pupils at Cotham School. Higgs, a pupil at the same Bristol institution, often wondered who Dirac was during morning assembly. It was his curiosity to discover what the great man had done that had led him to quantum field theory.

The quantum fields are ultimately what the world is made of.[6] Matter is composed of atoms. Atoms are composed of nuclei and electrons. Nuclei are made of protons and neutrons. Protons and neutrons are made of quarks (although, in 1964, this was still a very new idea). The quarks and electrons are made of fields. And as far as we know, the quantum fields are the bottom rung of nature's ladder.

A field is simply something that has a value at each point in space–time. It could be a number such as a temperature, or it could be a number with an associated direction such as a wind speed. Or it could be something altogether more complicated. Each fundamental particle has an associated field. There is an electron field, a photon field, an up-quark field, and so on. Such fields jitter because it is the nature of quantum things that they are inherently restless. And if a particular field is jiggled enough – in other words, if sufficient energy is injected into it – a ripple propagates through it, and this is a particle. A good visual image is of a wind disturbance propagating through a field of wheat. A ripple in the electron field is an electron, a ripple in the electromagnetic field a photon, and so on. The twist is that a quantum field cannot be rippled at any arbitrary rate but only at certain discrete frequencies, or energies. Its oscillations are said to be 'quantised'.

In this way, quantum field theory unifies the two puzzling and seemingly mutually exclusive properties of the subatomic world that were discovered in the early twentieth century: the ability of atoms and their constituents to behave both as particles and as waves.

One of the most remarkable features of the quantum field theory of the electron, which left a big impression on Higgs and many others, is that it is possible to deduce from it, almost trivially, the existence of the electromagnetic force, as described by James Clerk Maxwell in 1862. The key is 'symmetry', a property of any entity that remains unchanged when something is done to it. For instance, a circle has rotational symmetry because, when it is rotated about its centre, it remains the same. A square remains unchanged only if it is rotated by a quarter of a turn or by several quarter turns in succession.

In 1918, the German mathematician Emmy Noether, who was described by Einstein as the most important woman in the history of mathematics, proved a powerful theorem which relates to this. Whenever there is a symmetry, there exists a corresponding conservation law, which dictates that a certain physical quantity can neither be created nor destroyed. For instance, the fact that it makes no difference to the result of an experiment whether it is done today or next week – so-called 'time-translational symmetry' – corresponds to the 'law of conservation of energy', which maintains that energy can neither be created nor destroyed. And Noether's theorem has profound consequences for the quantum field theory of the electron, too, because transformations of its equations that do not change any of its observable consequences are also possible.

An electron is described by a 'wave function'. This spreads through space according to the Dirac equation, and the probability of finding the electron at any location is given by the square of the height of the wave (or, strictly speaking, the square of its 'amplitude', its maximum excursion from the zero level). The wave function has a 'phase', which describes where it happens to be in its undulating pattern. It turns out that changing the phase by the same amount everywhere – or, in the jargon, multiplying the wave function by a 'phase factor' – simply moves the peaks and troughs of the wave along and does not change anything observable, such as the probability of finding the electron at any particular location. The existence of this symmetry, according to Noether's theorem, must necessarily correspond to a conservation law. And it does: the law of the conservation of electric charge, which states that charge cannot be created or destroyed.

Noether's theorem applies to a change made everywhere at once that has no observable consequence, but such a 'global' symmetry is only one type of possible symmetry. There is another, far more restrictive type, in which the phase of the wave function is not changed by the same amount everywhere but by a different amount at every location of space and time. It might seem ridiculous to expect such changes to have no observable consequences and for the electron wave function to exhibit such a 'local symmetry', but it is not.

Imagine a billiard ball travelling across a billiard table in a straight line.[7] Raising the table vertically, whether by one metre or ten metres, does not change the trajectory of the billiard ball, which is dictated by Newton's laws of motion. But there is a hidden assumption here: that all parts of the billiard table can be raised simultaneously. While this is certainly true for a common-or-garden table, imagine a cosmic-scale billiard table, ten light years across. In this case, it would be impossible to make a change to all parts of the table simultaneously because, according to Einstein, nothing can travel faster than the speed of light. Distant locations will react to a change in the table's height later than nearby ones. In fact, it would be impossible for a change to the near side of the table to be 'noticed' by the far side before ten years have elapsed. In general, if an attempt is made to change the height of the billiard table, a step in height travels across its surface, causing the table to have a different height at different locations and times. This is the best that is possible in an Einsteinian universe.

Here is the point. We can still hope for the laws of physics to be the same everywhere, so that a billiard ball continues to follow the straight-line trajectory required by Newton's laws of motion. However, since the surface of the billiard table is no longer flat, this can happen only if at every location and at every time the billiard ball experiences a 'force' that precisely compensates for the uneven terrain.

The height of the table is a simple example of what physicists call a 'gauge', a term coined by the German physicist Hermann Weyl in 1929. The insistence that the laws of physics remain the same when the gauge is changed continuously from place to place and from time to time is known as 'local gauge invariance'. As the example of the billiard table shows, maintaining gauge invariance locally requires the existence of a compensating force. This is the key thing.

In the case of the electron, maintaining gauge invariance means insisting that there should be no observable consequences from continuously changing the phase of the electron wave function from location to location and from time to time. The phase of the electron wave function is obviously a more abstract mathematical thing than the height of a billiard table, but just as in that example, maintaining gauge invariance requires the existence of a compensating force. Remarkably, the force turns out to be the electromagnetic force described by Maxwell in the nineteenth century.

So the electromagnetic force – with its bewildering array of phenomena – is nothing more than an unavoidable consequence of local gauge invariance. Basically, the electromagnetic field exists so that when electric charges rearrange themselves at one location in space and time, news of this is communicated to other locations so local gauge invariance can be maintained. That news is carried by the electromagnetic field, which is composed of photons – ripples in the electromagnetic field.

Remarkably, even if we knew nothing about electricity, magnetism and photons but we knew about the gauge principle, we would be able to deduce that all these things exist simply to enforce local gauge invariance of the electron wave function. This extraordinary discovery was made in the 1950s by Julian Schwinger, one of the pioneers of quantum electrodynamics, the quantum theory of the electromagnetic force. It is so striking that it is natural to wonder whether it might be a universal principle. Could the necessity to enforce local gauge invariance be the reason for the existence of not just the electromagnetic force but all of nature's fundamental forces?

Besides the electromagnetic force, nature's fundamental particles are glued together by three other basic forces. The most familiar, gravity, was described by Einstein's general theory of relativity. Like electromagnetism, it is underpinned by a symmetry principle: the equations which reveal how gravity – that is, the curvature of space–time – depends on the distribution of energy have the same mathematical form for everyone, no matter what their motion or 'co-ordinate' system. Even Einstein's earlier 'special' theory of relativity arises from the insistence that the speed of light must appear the same to all observers moving at constant velocity relative to each other. In fact, Einstein was the first person to realise the importance of symmetry in underpinning the fundamental laws of nature. 'Nature seems to take pleasure in exploiting all possible symmetries for her fundamental laws, like a painter eager to use all the most splendid colours on her palette,' says Italian physicist Gian Francesco Giudice.[8]

But leaving aside the force of gravity, which for various reasons nobody has any idea how to express as a quantum field theory, there are two other fundamental forces, the strong and weak nuclear forces, which operate only within the ultra-tiny realm of the atomic nucleus.

The first person to seriously think that the gauge principle was the key to understanding not just the electromagnetic force but the strong and weak forces was the Chinese physicist Chen Ning Yang. 'The only true voyage of discovery', wrote Marcel Proust, 'consists not in seeing new landscapes but in having new eyes.' In the 1950s, Yang and his collaborator, the American physicist Robert Mills, opened new eyes on the world and wrote down an equation that a quantum field would have to obey to enforce a more generic local gauge symmetry on a wave function.

The Yang–Mills equation revealed that the electromagnetic field is the simplest possible 'gauge field'; not only is it transmitted by a single 'gauge particle', but that particle – the photon – carries no electric charge. Since electric charge is essential for a particle to interact with the electromagnetic field, the photon is immune to the electromagnetic force.

However, in the more complex gauge fields permitted to exist by the Yang–Mills equation, the force carriers do carry 'charges'. These analogues of electric charge, which can similarly be neither created nor destroyed, cause the force carriers to feel the gauge field. In the case of the strong force, for instance, such particles interact not only with the field but with each other, in complex ways that are difficult to predict. This hampered early attempts of physicists to show that the strong force is a consequence of local gauge invariance, as did their mistaken belief that the force principally operates between protons and neutrons. As became clear only in the late 1960s, protons and neutrons are composite particles, and it is their constituent quarks that are glued together by the strong force.

But those pursuing the idea that the fundamental forces are a consequence of local gauge invariance faced an even bigger obstacle. The quantum field theory of the electron, which originated with Paul Dirac in the 1930s, had been plagued by the equations 'blowing up' and making nonsensical predictions. It had been possible to remove such 'infinities' with a mathematical trick, but the problem was that this 'renormalisation' worked only if the force carriers were massless. Although this was the case for the electromagnetic field, whose gauge particle is the photon, it did not appear to be true for the strong and weak forces. The hint was their short range.

In the quantum picture, a force is viewed as arising from a kind of submicroscopic game of tennis. Force carriers are batted back and forth between particles, causing the particles to recoil from each other. Although this picture conveys the general idea, like many scientific analogies it is not perfect, since it explains the origin of a repulsive force but not an attractive one.

Because the laws that orchestrate the submicroscopic world are very different from those that govern the large-scale everyday world, the force-carrying particles such as the photon have a very special character. As mentioned before, one of the cornerstones of physics is that energy can be neither created nor destroyed, but is merely transformed from one form into another. In the quantum world, however, there is a twist. Energy can be conjured from nothing – strictly speaking, the vacuum – and nature will turn a blind eye as long as it is paid back quickly.

In 1905, Einstein discovered that mass is a form of energy – the most concentrated form possible – so the energy that is conjured from the vacuum can become the mass-energy of subatomic particles. Such particles, because of their fleeting existence, are known as 'virtual' particles, to distinguish them from 'real' ones. It turns out that the greater the energy borrowed from the vacuum, the faster it must be paid back. Consequently, the more massive a virtual particle, the more fleeting its existence and the shorter the distance it is able to travel before disappearing back into the vacuum.

Because the photon has no 'rest mass', it takes very little energy to create one, so there is very little energy to be paid back. This means it can exist for a very long time and reach the farthest corners of the universe, which is why the electromagnetic force has an infinite range. However – and this is the key point – the intimate connection between the range of a force and the mass of its force carriers suggests that the strong and weak forces, because of their ultra-short range, are transmitted by massive force carriers. (This logic turns out to be wrong for the strong force; nature, as will later become clear, has found another way of making it short-range.)

The problem is that the only quantum field theories that are devoid of catastrophic infinities are local gauge theories – something proved by the Dutch physicist Gerard 't Hooft in 1971 – but this essential feature exists only if the gauge particles are massless. It is lost the instant massive force carriers are introduced. This was one of the reasons why quantum field theory was not in favour in the early 1960s.

By July 1964, the problem of keeping all the desirable properties of a local gauge theory while also having massive force-carrying particles was something that had been occupying Higgs' mind for several years. What if the force carriers were intrinsically massless, he wondered, but were given masses by some external process? Might that square the circle of having massive particles but retaining a renormalisable local gauge theory? Higgs imagined space being filled with a previously unsuspected invisible field, which, like the water in a swimming pool, resists motion through it.

At first glance, the swimming pool idea chimes perfectly with our experience of mass. A body with a big mass, such as a fridge, is hard to budge – it resists attempts to change its motion. This might plausibly be because there is an invisible medium pushing back. However, the analogy is imperfect – a basic feature of the world, and a foundation stone of Einstein's special theory of relativity, is that no experiment can reveal whether you are moving at constant speed or stationary. If two people are playing catch on a train and the windows are blacked out and the train is vibration-free, the ball will loop back and forth between them just as it would if they were standing beside the track. They will be unable to tell from the ball's motion whether they are in motion or not.

However, if Higgs' ubiquitous space-filling field was exactly like the water in a swimming pool, it would be possible to distinguish between a body moving through it and one that was stationary, contradicting relativity. Instead, the Higgs field had to appear stationary to every body in the universe, irrespective of its motion.* This property of being, in the jargon, 'Lorentz invariant', is true only of 'scalar fields' that, like temperature and the height of a billiard table, are characterised by a simple number at each point in space.

The twist is that the Higgs field resists particles even though they are always stationary with respect to it. A more accurate statement, therefore, is that the field simply interacts with them. It is this interaction that causes intrinsically massless particles to have masses, with their precise mass depending on how strongly each particle interacts with the field.

A field whose energy is non-zero everywhere in space was something entirely new. In the absence of mass there is no gravitational field, and in the absence of electric charge there is no electromagnetic field. But Higgs imagined a field which existed in otherwise empty space and had no source. The field had the same energy everywhere, and it was this sameness that explained

---

* How can something appear the same to everyone, no matter what their velocity? Imagine a rainbow. The colours are known to be a measure of the distance between successive peaks of light waves, and there are waves with both shorter and longer 'wavelengths' than visible light. By convention, a rainbow is said to contain seven colours. Think of them as numbered from one, representing the longest wavelength, to seven, denoting the shortest. It turns out that an infinite number of 'colours' are possible. Imagine labelling them '–[infinity]' to '+[infinity]'. Now imagine that they all exist in space. If you fly through them at constant velocity, all the waves will appear scrunched up, or 'Doppler shifted', so one will become two, two will become three, and so on. The result of shifting all the colours in this way, however, will still be a set of colours that span the range from –[infinity] to +[infinity]. Consequently, it will be impossible to tell that you are moving with respect to the light. In a sense, all colours do exist in space because, according to quantum theory, every vibration 'mode' of the electromagnetic field must contain a minimum amount of energy. And what is true of the electromagnetic field is true of every field, including the Higgs field.

why it had never been noticed before: we are immersed in it, just as we are immersed in the air we breathe.

Given the opportunity, everything minimises its 'potential energy'. For instance, a ball will roll to the foot of a hill, where its 'gravitational potential energy' is lowest. It had always been assumed that the vacuum was the lowest energy state of the universe; it was where, as surely as a ball rolling to the bottom of a hill, the universe would end up. However, Higgs was suggesting this is wrong and that the lowest-energy state of the universe is actually a vacuum filled with the Higgs field, with a non-zero energy everywhere.

But arbitrarily introducing into a theory a field which fills all of space in order to give masses to the force carriers is as artificial as inserting their masses by hand. It was also guaranteed to wreck local gauge invariance, which was an essential requirement for a theory without monstrous infinities. What Higgs needed was a way to introduce his field in a natural way, and he had seen how to do it.

A gauge theory in which all the force carriers are massless is elegant and symmetric; after all, the masses of all the particles are exactly the same. On the other hand, a theory in which the particles have masses that may not be the same is messy and less symmetric – physicists say its symmetry is 'broken'.

Examples of a symmetry breaking naturally are easy to find in the everyday world. Imagine a pencil balanced vertically on its sharpened end. It is perfectly symmetric – it looks the same from every direction. However, if it is buffeted by a draught of air, it may fall pointing north, southwest or in any other direction. Its orientation with respect to the vertical is no longer symmetric. This illustrates that although the fundamental laws of physics may be symmetric – in this case, the force of gravity points downwards and favours no direction of the compass – the outcome of those laws may nevertheless be asymmetric.

The field Higgs had in mind was one that was perfectly symmetric – in effect, 'switched off' and incapable of upsetting local gauge invariance – but whose symmetry was spontaneously broken, causing it to switch on and interact with the gauge force carriers to give them masses. The simplest scheme he could imagine involved the energy of the field being given by the height of a marble on a 'potential' the shape of a sombrero hat. Initially – and this would have been in the super-high-energy conditions of the Big Bang – the ball would have been in a perfectly symmetric state on the central peak of the hat. However, as the universe cooled down towards its present-day low-energy state, the marble would have rolled into the gutter of the hat, picking a direction and breaking the symmetry.

An explanation of how the interaction of the massless force carriers with Higgs' spontaneously broken field generates their masses will have to wait; there is a more immediate and pressing problem with Higgs' scheme, which other physicists were aware of and which was another reason why quantum field theories were out of favour in the early 1960s.

Think of the sombrero hat again. The marble characterising the energy of the Higgs field could end up anywhere around the rim, with each location corresponding to a state of the Higgs field. An atom can change from one energy state to another by emitting or absorbing a photon with an energy equal to the energy difference between the states. However, in the case of the Higgs field, each state around the rim of the sombrero is at exactly the same height and so has exactly the same energy. Changing from one state to any other takes no energy, which means it corresponds to a particle with zero mass – such a particle is known as a 'Goldstone boson'.

The British physicist Jeffrey Goldstone had discovered that such particles are an unavoidable consequence of spontaneously breaking the symmetry of a scalar field. The problem was that having zero mass, they should be very easy to create and so should have revealed themselves long ago in physicists' experiments. But not a single Goldstone boson has ever come to light. Physicists had taken the absence of Goldstone bosons as proof that quantum field theories that require spontaneous symmetry breaking to make contact with the real world are a theoretical dead end.

Yet the idea that local gauge symmetry might spawn the fundamental forces was hugely appealing to Higgs. It was elegant, beautiful and it felt right – he just could not let it go. For years, he had mulled it over in his mind, and then he had his breakthrough. Three weeks earlier, he had sent his paper to *Physics Letters*, in which he set out a striking result. Higgs' breakthrough was to find that Goldstone bosons go away if a quantum field theory is locally gauge invariant – that is, if there are also gauge force carriers. To appreciate why, it is necessary to know about a key distinction between massless and massive particles.

A subatomic particle, as pointed out, is simply a wave propagating through a quantum field, much like a wind rippling through a field of wheat. The world in which we live has three dimensions of space, so it seems obvious that a wave representing a particle can oscillate in three mutually perpendicular directions. This intuition is correct for a particle which has a mass, but it is not true for a massless particle such as a photon, which travels at the speed of light.

A photon is associated with an electromagnetic wave, whose electric and magnetic fields oscillate in a plane perpendicular to the direction of the wave's travel. In addition to these two 'transverse' oscillations, it might be expected that the wave would also oscillate in its direction of motion, but such a 'longitudinal' wave would necessarily alternate between moving slower than the speed of light and faster than the speed of light. This is impossible because, according to Einstein, the speed of light is the ultimate cosmic speed limit. The upshot is that while a massive particle has three independent ways of oscillating, a massless one has only two.

Higgs' breakthrough was to realise that the Goldstone bosons go away miraculously in a theory with massless gauge particles. They are, in effect, 'swallowed' by the force carriers. Not only does the 'Higgs mechanism' get rid of the troublesome Goldstone bosons but, in being swallowed, they endow the massless gauge particles with a third way of oscillating, giving them masses.[9]

It should be pointed out that this mechanism for acquiring mass is different from the one previously described in which particles interact with the ubiquitous Higgs field, encountering resistance like a swimmer ploughing through a swimming pool. Indeed, nature has seen fit to provide two separate mechanisms for endowing particles with mass. One, which gives mass to nature's force carriers, involves spontaneous symmetry breaking; the other, which gives mass to nature's building blocks – the quarks and leptons – involves a more straightforward interaction with the Higgs field.

The Higgs mechanism is a miracle that kills two birds with one stone: it simultaneously gets rid of the Goldstone bosons and endows the gauge force carriers with masses. The role of Goldstone bosons, according to Steven Weinberg, is changed 'from that of unwanted intruders to that of welcome friends'.[10] Beneath the mass-inducing cloak of the Goldstone bosons, the force carriers remain massless and, consequently, they continue to be described by a renormalisable, infinity-free local gauge theory.

o   o   o

Higgs had spelled all this out in two short papers in the summer of 1964. In the first, he showed how it was possible to get rid of Goldstone bosons in a quantum field theory, as long as gauge bosons are also present. In his second paper, he outlined how the gauge bosons acquire masses, by cannibalising the Goldstone bosons.[11] But as he sat in his office, with his rejected second paper sitting on the desk before him, he still faced the problem of what to add to ensure it would be accepted for publication. Maybe a physical consequence of his idea?

Every quantum field has a particle or particles associated with it because every field can be rippled, and a ripple propagating through the field is all a particle actually is. Higgs knew his symmetry-breaking field would be no exception.

He thought again about the sombrero hat potential that governed his field. The Goldstone bosons arose because the marble could oscillate around the rim of the sombrero, but that was not the only type of oscillation that was possible. The marble could also oscillate in a radial direction, up and down the valley formed by the rim of the sombrero. Such an oscillation would require a minimum amount of energy to excite it, and energy, according to Einstein, has an equivalent mass. If sufficient energy were pumped into a small area of space, it would be possible to create such a particle: a ripple in the Higgs field.

Higgs' annoyance at the rejection of his second paper by *Physics Letters* had not simply been because its editor failed to appreciate the importance of his work. Jacques Prentki had made things even worse by suggesting not that he resubmit his revised paper to *Physics Letters*, which ran all the papers it received past independent scientific 'referees', but rather to *Il Nuovo Cimento*, the Italian journal which did not bother with referees at all. Prentki seemed to think the paper was irrelevant and worthless, and that stung.[12]

Consequently, Higgs was damned if he was going to follow Prentki's advice. He resolved instead to send the revised paper not to *Physics Letters* but to its American rival, *Physical Review Letters*. However, first he needed to add something, and at last he knew what. He picked up a pen and supplemented his paper with a final, two-sentence paragraph. In the first sentence, he wrote, 'It is worth noting that an essential feature of the type of theory which has been described in this note is the prediction of incomplete multiplets of scalar and vector bosons.' What Higgs meant in this highly technical statement was that there would be a particle left over from the symmetry-breaking process. A Goldstone boson with mass, a hitherto unexpected fundamental particle.

What Higgs did not know was that five other physicists had come to exactly the same conclusion at pretty much the same time. In London there was a 'gang of three' consisting of Tom Kibble, Gerry Guralnik and Dick Hagen, and in Brussels there was a 'gang of two' consisting of Robert Brout and François Englert. Higgs, as he would later refer to himself, was the 'gang of one'.

As Higgs took his amended paper to the departmental secretary to be retyped, he felt a warm glow of satisfaction. He had no idea how it might fit into the framework of particle physics, but he was sure it was important. He also did not have the slightest inkling that the sentence he had added would earn him immortality; in fact, it would win him the Nobel Prize.

o   o   o

Higgs' work created no sensation and grabbed no headlines. As Tom Kibble, one of the 'gang of six', would later say, 'Our work was greeted with deafening silence.' In devising his mechanism for supplying mass to the gauge force carriers, Higgs had been hoping to make sense of the strong nuclear force that binds together the components of the atomic nucleus, but that problem was not ripe for the picking. The strong force was believed to act principally between protons and neutrons, but these were composite particles and the strong force glued together their constituent quarks, something that was only recognised by the physicists Murray Gell-Mann and George Zweig in the mid-1960s.

If this was not enough of an issue in trying to understand the strong force, there was another fundamental difficulty: it turned out that the short range of the force was not, as might be expected, a result of its force carriers being massive and therefore being conjured only fleetingly from the vacuum. The 'gluons' are massless, so the Higgs mechanism, as far as the strong force is concerned, is irrelevant. Nature has thrown us a curve ball and has chosen to use an entirely different mechanism for giving the strong force its short range.

Just as the restoring force in a piece of elastic becomes stronger the more it is stretched, so the force between two quarks gets stronger the further they are pulled apart. In fact, so much energy must be put into separating a pair of quarks that it conjures into existence the mass-energy of a quark–anti-quark pair, which in turn have to be pulled apart, conjuring another quark–antiquark pair into existence . . . Since it is never possible to completely separate two quarks, the strong force never gets the opportunity to operate over anything but an ultra-short range.

Local gauge symmetry is maintained in the case of the strong force for the simple reason that the gauge force carriers – the gluons – retain their masslessness. Along with the quarks, they are imprisoned inside protons and neutrons, and permanently concealed from view. Whereas the massless symmetry of the weak force is hidden by symmetry breaking, the massless symmetry of the strong force is hidden by 'quarks confinement'.

The upshot of all this is that Higgs, in inventing a mechanism for giving mass to force-carrying particles, had been thinking about the wrong force. The right force was in fact the weak nuclear force, something recognised in the late 1960s by Steven Weinberg in the US and Abdus Salam in the UK. The pair were engaged in a struggle to show that the weak and electromagnetic forces have a common origin.

Recall that the great triumph of nineteenth-century physics was the discovery by James Clerk Maxwell that the electric and magnetic forces have a common origin.[13] But in that case, there had been strong hints that they were related. Hans Christian Ørsted had shown that a changing electric field creates a magnetic field, and Michael Faraday had shown that a changing magnetic field generates an electric field. But with the electromagnetic force and the weak force, it was not at all obvious that there was a connection.

The electromagnetic force has an infinite reach, whereas the weak force has a range barely of 1 per cent of the diameter of a proton. And whereas the electromagnetic force merely moves charged particles around in space, the weak force can shift electric charge between particles, magically transforming one kind of particle into another – for instance, a neutron into a proton in the process of radioactive beta decay (though what actually happens is the weak force turns one 'flavour' of quark – a down-quark – inside a neutron into another flavour of quark, an up-quark).

The weak force is critically important in the Sun because nuclear reactions which require the weak force are rare. (In the quantum world, weak is synonymous with infrequent.) The rareness of the first step in the chain of sunlight-generating nuclear reactions is the reason the Sun will use up its fuel gradually over ten billion years or so and not squander it explosively in one go. This has enabled the Sun to shine steadily for the billions of years necessary for the evolution of complex life. And if this is not enough to be thankful for, the weak force is also of key importance in the nuclear processes inside massive stars that built up elements such as carbon, oxygen and iron that have been crucial to life on Earth.*

The weak force appears so different from the electromagnetic force that the claim that they have a common origin is a brave one. How in the world can rainbows and radioactive decay be aspects of the same fundamental phenomenon? But

---

* Another aspect of the weak force is that it affects only particles that spin one way. Think for a minute how weird this is. Imagine there is a Category Five hurricane and lots of couples are dancing in it. Those spinning clockwise are instantly blown away, while those spinning anticlockwise are unaffected. This remarkable – in fact, scarcely believable – aspect of the weak force – that it violates so-called left–right symmetry – was discovered by Chinese–American physicist Chien-Shiung Wu in 1956.

this is exactly what Schwinger suggested in 1956.[14] In the 1960s, Weinberg, Salam and others set out to demonstrate that he was right. It was in the 'unification' of the electromagnetic and weak force into the 'electroweak' force that Higgs' idea proved its worth.

The weak force's short range is indeed explained by its gauge force carriers having large masses – almost one hundred times the mass of a proton. Because weak-induced beta decay adds positive electric charge to a neutron to make a proton, there must be a force carrier with a positive charge: this is the W+. And because beta decay can work in reverse, adding a negative charge to a proton to make a neutron, a W– must also exist. The existence of the W+ and W– was, in fact, predicted by Schwinger. For technical reasons, there must also exist a weak force carrier with no electric charge: this is the Z0, predicted by American physicist Sheldon Glashow in 1960. Together, the W+, W– and Z0 are known as the 'weak vector bosons'.

A quick word on what a boson is. In nature, particles can carry an intrinsic, or 'quantum', spin, and it can only be a whole-number multiple of a basic spin (such as 0 or 1) or a half-integer multiple of that spin (such as ½ or ³⁄₂). The first type of particles are known as bosons and the latter are fermions. Force-carrying particles such as photons and gluons are bosons, while building-block particles such as quarks and electrons are fermions. There is a profound connection between the spin of particles and how they behave en masse. According to the 'spin-statistics theorem', two bosons with identical properties can be in the same place at the same time, but two fermions cannot. This is why photons are happy to travel together in their countless quadrillions in a laser beam, whereas electrons do their best to avoid each other. It is this unsociability of electrons that explains why they occupy separate orbits in atoms, thus making matter extended and solid bodies possible.

In the theory of Weinberg and Salam, the electromagnetic force exists to maintain a local gauge symmetry known as U(1); basically, to keep a number – the 'complex' phase of a quantum wave of an electron – the same at every point in space–time. The weak force exists to maintain a slightly more complex symmetry known as SU(2), which involves a 2 × 2 matrix that is similar to, but not quite the same as, the one used by Paul Dirac in his celebrated equation (see chapter 'Mirror, mirror on the wall'). Together, the two symmetries are known as U(1) × SU(2). They mix with each other, as was also realised by Glashow, so the electroweak force carriers turn out to be the photon and the Z0, which arise from mixtures of the two symmetries, and the W− and W+. The Z0 is nothing more than a massive photon, which we can think of as 'heavy light'.

Since the gauge force carriers exist to enforce local gauge symmetry, they are all massless. This would have been the case in the super-high-energy conditions in the earliest moment of the Big Bang. Enter the Higgs mechanism, and the Higgs field in this situation is a little more complicated than the one illustrated with the sombrero hat. It is a so-called SU(2) state with four components, which spawn four Goldstone bosons, or 'Higgses'. Three are cannibalised by the W+, W− and Z0, in the process giving them masses. (The photon does not participate in this game and remains massless.) That leaves one Higgs boson with intrinsic mass – this is the 'left-over' particle whose existence was predicted by Peter Higgs in August 1964.

By July 2012, strong evidence of all the fundamental particles of the 'Standard Model' – the quantum field theory of the three non-gravitational forces – had been found in experiments.[15] They include six quarks, known as up, down, strange, charm, bottom and top; six leptons, known as the electron, electron neutrino, muon, muon neutrino, tau and tau neutrino; and the twelve force carriers. Of these, the photon mediates the electromagnetic force; the W+, W– and Z0 the weak force; and eight gluons the strong force.*

It is not quite true that all the particles of the Model had been found. All of them had been found, except one: the Higgs.

Methodist Central Hall, London, 4 July 2012

On the screen at the back of the stage, the press conference at CERN was starting. The laboratory's main auditorium was rammed full, with its steeply banked seats even more jam-packed than the cavernous Methodist Central Hall in London. Butterworth was already fielding questions from eager journalists, and to his frustration, was only able to glance intermittently at the giant screen.

At CERN, the spokespersons for the ATLAS and CMS experiments, Fabiola Gianotti and Joe Incandela, were setting the scene. The Large Hadron Collider is the most complex machine ever built. It fills a twenty-seven-kilometre circular tunnel beneath the Swiss–French border that is as long as the

* The building blocks of matter – the quarks and leptons – are fermions, whereas the force-carrying particles that glue them together are bosons. All normal matter is built of only four of these particles – the up- and down-quark and the electron and electron neutrino (a proton in an atomic nucleus consists of two ups and one down, while a neutron consists of two downs and one up). The other quarks and leptons are merely heavier versions of these. It is a complete mystery why nature has chosen to triplicate its basic building blocks in this way.

Circle Line on the London Underground. It was previously occupied by the Large Electron–Positron collider (LEP).

LEP was limited in the collision energies it could reach because whenever electrons and positrons are accelerated – something that happened when their paths were bent into a circle by the powerful magnets around the LEP tunnel – they broadcast electromagnetic radiation, which sapped them of energy. Crucially, however, such 'synchrotron radiation' is more of a problem for light particles than heavier ones. In fact, it depends on the 'inverse fourth power' of the mass of a particle and so, for protons, which are about two thousand times heavier than electrons, it is about ten trillion times less serious than for their lighter cousins. This is the reason why, when LEP was ripped out of the tunnel, it was replaced with a proton–proton collider: the LHC. (A 'hadron', incidentally, is any particle that experiences the strong nuclear force.)

Constraining the super-high-energy protons to circle their subterranean race track one hundred metres beneath the surface of Switzerland and France requires bending their paths with the strongest possible electromagnets, which requires their coils to be powered with the largest possible electric currents – 12,000 amps in the case of the LHC. Such high currents would normally generate huge amounts of heat, but the 1,232 bending magnets of the LHC, each fifteen metres long and weighing thirty-five tonnes, are formed from special 'superconducting' coils, which are cooled by liquid helium, the world's best refrigerant. At $-271.3°C$ – a mere 1.9 degrees above the lowest temperature possible – the coils offer no resistance to an electric current and so dissipate no heat while they remain superconducting. However, in tests shortly after 10 September 2008, when the LHC's proton beams were first switched on, a connection between two magnets lost its superconductivity. This led to a spark that punctured the twenty-seven-kilometre-long cooling vessel – the biggest refrigerator ever built – causing an explosion as the escaping liquid helium rapidly turned to gas, leading to extensive damage to 750 metres of the magnet ring. The accident set the programme back by more than a year.

But since restarting in November 2009, the LHC had worked without a hitch. At ATLAS and CMS, collisions between protons travelling at 99.9999991 per cent of the speed of light recreate, for the briefest of instants, conditions that last existed a hundredth of a billionth of a second after the birth of the universe, when the temperature of the Big Bang fireball was about ten million billion degrees. From the energy of the collisions – not strictly speaking between protons but between their constituent quarks and gluons – are conjured 'jets' of quarks and gluons. They spawn exotic particles, which live for the tiniest slivers of time before transforming into yet more subatomic debris. The hardware and software is primed to filter out the majority of events from this bewildering subatomic mayhem, leaving only the rarest of rare signatures the physicists are looking for: the signature of the Higgs.

If created, the elusive subatomic particle is expected to survive for too short a time to be directly detected; the trick is to look for particles into which it decays, which are distinguishable from particles generated by a myriad of other confusing 'background' processes. ATLAS looked for rare pairs of photons generated by the decay of W+ and W– particles, in turn spawned by the Higgs. CMS looked for rare pairs of Z0s, also from the decay of a Higgs. The physicists working on the ATLAS and CMS experiments were kept in the dark about each other's progress as far as possible. What CERN wanted more than anything was one experimental result confirmed by a second, entirely independent, result.

Butterworth knew roughly what Gianotti would be saying because he had been at CERN's Salle Curie conference room the day before, when she rehearsed her presentation in front of her colleagues. Nevertheless, he was keen not to be distracted when her presentation reached its crescendo. Thankfully, however, the mounting excitement in Switzerland had spread to London. The journalists in Methodist Central Hall had fallen silent and all eyes were on the giant screen.

Gianotti was showing a graph. It showed a hump at 126GeV, roughly 126 times the energy required to create a proton. It was exactly what would be expected if the decay products seen by ATLAS and CMS came from a particle which existed for the most fleeting of instances. A new particle, hitherto unknown to science. Gianotti said the magic words: 'For both experiments, the discovery has now reached the "five-sigma" level of confidence.'*

At CERN, there was pandemonium. The audience, which for the past half-hour had listened patiently to the technicalities, erupted into raucous applause and cheering. The TV picture panned to a flushed and smiling Peter Higgs, who had been invited to attend and was sitting squeezed in the middle of the auditorium. People all around him were congratulating him and reaching across to shake his hand. A modest man of

---

* Sigma is a probability measure: the larger its value, the more certain physicists are that a result is real and not just a random fluke in their data. At a 'five-sigma' level of confidence, physicists know there is a one in two million chance that nature has hoodwinked them, which is why it counts as a 'discovery'.

eighty-three, he looked slightly bewildered. He was pushing up his glasses and appeared to be wiping away a tear. François Englert, a member of the 'gang of two', was also there. In a year's time, the two men would share the 2013 Nobel Prize in Physics.[16]

In Methodist Central Hall in London there was also pandemonium. People were on their feet, cheering. Butterworth had entirely forgotten his feelings when he arrived at the building earlier. This was a piece of history, and even though he was 750 kilometres from Geneva, he was a part of it. He had thought he was working in an esoteric field of physics and that nobody in the wider world really cared, but the evidence was all around him that he was wrong. Everyone, no matter how much physics they understood, realised that this was a key moment in the history of science. A key moment in the history of the human race.

In the summer of 1965, a shy, modest man in an office at the University of Edinburgh added two sentences to a paper that had been rejected for publication, predicting the existence of a hitherto unexpected massive particle. Now, almost four decades later and at a cost of some five billion euros to construct and run the most complex machine ever built, there it was.[17] Or was it?

'Is this the Higgs, Professor Butterworth? Is this really the Higgs?' came the question.

'We have found a new and real particle,' Butterworth replied, choosing his words carefully. 'It's consistent with the Higgs.'

'But is it the Higgs?'

'We think it's the Higgs, but we need to do some more work to be sure. What we know is that it is a new particle. That's what's so exciting: a new particle!'

But that was not enough for the journalists. The question kept coming back. It was relentless, and Butterworth found it both funny and frustrating. 'Have you found the Higgs?'

Butterworth was unwilling to go there yet. Just because something looked like the Higgs particle did not mean it was the Higgs particle. Butterworth and his colleagues needed to measure the properties of the new particle – its quantum spin and the precise details of its decay – to see whether it was the Higgs boson, as described by the Standard Model. But in his bones, in his heart of hearts, he knew. It looked like the Higgs. It smelled like the Higgs. At long last, they had found the Higgs.

o o o

The discovery of the Higgs was monumental. It is the last jig-saw piece of the Standard Model, the high point of 350 years of science. We have identified the fundamental building blocks of the universe and understand the forces that bind them together. Everything exists – you and me, digestive biscuits, snails, soap operas, giraffes, stars and galaxies – to enforce local gauge symmetry, one simple principle from which everything arises.

Nobody knows why nature has such a strong desire to enforce local gauge invariance. In the words of the great Italian physicist Enrico Fermi, 'Before I came here I was confused about this subject. Having listened to your lecture I am still confused. But on a higher level.' But the discovery of the Higgs is a dramatic confirmation of the power of science – its central magic. That people can see things in the mathematical equations they have concocted to describe nature and then go out and find them in the real world. 'That equations written on paper can know nature, and that forty-eight years later experiments can prove this, is awesome,' says Frank Close. 'An overworked adjective but on this occasion justified.'[18]

The Higgs particle is unique in the Standard Model. It is the only elementary boson that is not a gauge particle – the only boson that is not a force carrier. It has no electric charge and no quantum spin. In fact, it is the first spin-0 particle ever discovered. The carriers of the electromagnetic, weak and strong forces all have spin 1, whereas the 'graviton', the hypothetical carrier of the gravitational force, is expected to have spin 2.

The Higgs, weighing in at 126 times the mass of the proton, is the heaviest subatomic particle ever detected. Being so heavy, it interacts most frequently with other heavy particles such as the top- and bottom-quark and the heavy tau lepton, and it appears to behave exactly as predicted by theory. There is no reason to believe that it will not interact as predicted with the lighter quarks and leptons. However, such interactions are rarer and it will take a lot more data to confirm them, just as it will to observe the Higgs interacting with itself.

The Higgs is not the 'God particle', as it was dubbed by Leon Lederman, but, to hijack the words of the Indian novelist Arundhati Roy, it is 'the god of small things'.

But although the Higgs particle is important, it is merely a short-lived ripple of the Higgs field; its true significance is in confirming the existence of the field itself and in beginning to reveal its properties. The Higgs field was always the key thing, but it was a more esoteric entity to sell to the public than the prospect of discovering a new subatomic particle.

The Higgs field is something truly new. As mentioned earlier, every other field is zero in empty space. It may jitter a little because of quantum uncertainty, but it averages out to zero. The Higgs, however, is non-zero everywhere in space, and because it is ubiquitous, everything in the universe spends its life immersed in it. Until 4 July 2012, this was only a theoretical possibility. But now, because we have observed a ripple in the Higgs field – the Higgs boson – we know it is really there. And every fermion – every quark and lepton – interacts with it constantly. Like the W+, W– and Z0, they are intrinsically massless, but their mass depends on how strongly they interact with the Higgs field.[19] What for centuries people have called 'mass' is now known to be a consequence of the interaction between the fundamental particles and the Higgs field.

And what of the Higgs boson itself? Well, it gets its mass – wait for it – by interacting with itself!

There is a proviso here. Although two separate mechanisms give particles mass – the devouring of the Goldstone bosons in the case of the W+, W– and Z0, and the interaction with the Higgs field in the case of the fermions (and even the Higgs) – these mechanisms turn out to be responsible for only about 1 per cent of the mass of your body. This is because its major building blocks are quarks, and the lion's share of their mass is explained not by the Higgs but by Einstein's special theory of relativity. Inside the protons and neutrons of atomic nuclei, the quarks fly around at close to the speed of light and, as Einstein showed, bodies become more massive as they approach light speed.*

If the physicists had sold the LHC to politicians by saying it was going to find the reason for 1 per cent of mass, they probably would not have got very far. But that 1 per cent is of key importance because without it, the quarks and electrons in your body would be massless. This would mean they would travel at the speed of light and would be unable to settle into

---

* Because the strong force between quarks gets stronger the further apart they are, it follows that it gets weaker the closer together they are. Inside protons and neutrons, the force is so weak that the quarks behave like free particles. They are said to be 'asymptotically free'.

atoms. Everything would fly apart. Without the Higgs field, you and me, the stars and the galaxies, would not exist.

Of course, this is exactly the way it was in the earliest moments of the Big Bang. At the high energies that existed then, all particles were massless and travelled at the speed of light, interacting with each other in a completely different manner to the way in which they do in today's low-energy universe. 'The past is a foreign country: they do things differently there,' as novelist L. P. Hartley observed in *The Go-Between*. It is our great triumph as human beings to have discovered this. It was the switching on of the Higgs field that made everything we see around us possible.

The Higgs field, being ubiquitous, may play some other as-yet-unsuspected role in controlling the universe. But even if it does not, its mere existence shows us that scalar fields are possible. Such fields, as already mentioned, have the key property that they appear the same to everyone no matter what their velocity, and therefore do not conflict with the requirement of Einstein's special theory of relativity. The existence of the Higgs field raises the possibility that the universe may contain other scalar fields, and that these may explain some of its deeply puzzling features. For instance, the universe is believed to have gone through a brief phase of accelerated expansion, known as 'inflation', during its first split second of existence and, bizarrely, is undergoing a far weaker and more sustained form of accelerated expansion today, powered by mysterious 'dark energy'. Theorists suspect that the former phase was driven by a scalar field which exerted repulsive gravity known as the 'inflaton', and that the latter may also be driven by a scalar field.

The truth about the Higgs field, however, is that we know that it exists but we do not know its origin, why it has a non-zero average value in empty space or whether it is actually fundamental. Conceivably, it could be a composite of fields like protons and neutrons, which are made up of three separate quark fields. The hope among physicists is that as they learn more about the Higgs, they will get new physical insights because, although the Standard Model is a triumph, there is so much about it that is arbitrary and mysterious.

Physicists do not know, for instance, why the fundamental particles have the masses they have – why are top quarks roughly a trillion times heavier than neutrinos? – and why the fundamental forces have the relative strengths they have. We do not know why the electromagnetic force is an extraordinary ten thousand billion billion billion billion times stronger than the gravitational force. And why are there three families of quarks and three families of leptons, with each generation more massive than the one before? Even more seriously, there is no place in the Standard Model for gravity or for 'dark matter', which is known to outweigh the visible stuff in the stars and galaxies by a factor of six. The Standard Model is an approximation of a deeper theory. And it is that deeper theory that everyone is desperate to find.

9

The voice of space

If you ask me whether there are gravitational
waves or not, I must answer that I do not know.
But it is a highly interesting problem.
ALBERT EINSTEIN

Ladies and gentlemen, we have detected
gravitational waves. We did it!
DAVID REITZE, 11 FEBRUARY 2016

In a galaxy far, far away, at a time when the most complex
organism on Earth was a bacterium, two monster black holes
were locked in a death spiral. They whirled around each other
one last time. They kissed and coalesced. And, in that instant,
three times the mass of the Sun literally vanished. It reappeared
a moment later as a tsunami of tortured space–time, surging
outwards at the speed of light.

For a brief instant, the power in the 'gravitational waves' was
fifty times greater than the power radiated by all the stars in the
universe combined. In other words, had the black hole merger
created visible light rather than violent convulsions of space–time,
it would have shone fifty times brighter than the entire universe.

The gravitational waves spread outwards like concentric ripples on a pond. They buffeted a million galaxies. They jiggled a
million million million stars. They tickled planets and moons
and asteroids and comets without number.

On Earth, great tectonic plates bucked and spun about and
crashed together, rearing up into towering mountain ranges,

which were ground back down to nothing by wind and rain and ice. Life, which had been stalled at the single-celled stage for three billion years, made the improbable leap to multicellular organisms. Plants and animals proliferated, spreading across the face of the planet, extinguished repeatedly by impacting chunks of interplanetary rubble, only to rise again. The dinosaurs came and went. Ice spread down from the poles, before returning whence it came over and over, ebbing and flowing like a mile-deep white tide. An upright ape arose and left footprints in the volcanic ash of Laetoli in Tanzania, and whose descendants, barely an instant of geological time later, left bootprints in the dust of the Moon's Sea of Tranquility.

The ripples in space–time rolled onwards. They lapped at the outer shores of the Milky Way. They surged inward to the Orion Spiral Arm. They jostled the cloud of icy comets at the fringes of the solar system. They sped past the gas giant planets and their mega-moons and on towards the rocky planets that huddled close to the fires of the Sun. They tickled Mars, the Moon and the top of the Earth's atmosphere. And finally, after their immense 1.3-billion-year journey across space, they ran into something that had been patiently waiting for them.

Hanover, Germany, 14 September 2015

It was around midday when a 'ping' announced the arrival of an email. Marco Drago, sitting at his computer, did not immediately look to see what the email said because he was busy writing a scientific paper. Every day at this time, an email like this came in, and always it was routine and of no particular significance.

For Drago, Monday 14 September 2015 had started out as an ordinary day. On a sunny autumnal morning, he had left his flat in Nordstadt, a quiet district close to the centre of Hanover, and walked for ten minutes to the Max Planck Institute for Gravitational Physics. Employing about 200 people, the establishment consisted of a pair of modern rectangular buildings, separated by a roadway and connected by a glass corridor. Drago had arrived at his first-floor office at 9am, taken off his coat and sat down at his laptop to check the emails that had arrived overnight. About one thousand people were involved in the LIGO–Virgo collaboration, and time differences between their different countries meant that the electronic chatter was non-stop.

Drago had come to Germany the year before from the University of Trento near Verona. His postdoctoral job at what was informally known as the Albert Einstein Institute required him to monitor one of the algorithms that analysed the outputs of the two LIGO gravitational wave detectors in the US and the Virgo detector in Europe, and which extracted any 'trigger signals' that might conceivably be candidates for gravitational waves.

Drago saved what he had written of his paper, flipped to his email inbox and clicked on the alert. The Virgo detector, near Pisa, was not yet working, so there were just two attachments. One came from Livingston in Louisiana, and the other from Hanford in Washington state.

At Livingston was a four-kilometre 'ruler' made of laser light, and three thousand kilometres away in Hanford was an identical four-kilometre ruler made of laser light. When Drago clicked on the two attachments and displayed them on a split screen one above the other, he saw that, at 5.51am Eastern Standard Time, a shudder had gone through the Livingston ruler, and seven milliseconds – less than a hundredth of a second – later, an identical shudder had gone through the Hanford one. They were displayed as wiggly lines, proceeding from left to right, their up-and-down excursions mirroring perfectly the stretching and squeezing of the giant rulers. For about a tenth of a second, the wiggles became faster and more frantic before reaching a crescendo and abruptly dying away.

It was the unmistakable signature of a passing gravitational wave. A passing gravitational wave from a pair of merging black holes. But that was simply too ridiculous for Drago to accept. And not for a moment did he allow himself to believe that that was what he was seeing.

Although Advanced LIGO – the Laser Interferometric Gravitational Wave Observatory – had been operational for a month, it was still undergoing engineering upgrades to boost its sensitivity and was not scheduled to begin scientific work until four days later, on 18 September. The LIGO project had begun in the 1980s, but only now, in its 'Advanced' incarnation, was it approaching the sensitivity necessary to make a detection. What was the chance, thought Drago, that so soon after being switched on, it would pick up a gravitational wave? Next to zero.

This was not the only thing that set alarm bells ringing in Drago's mind. Gravitational waves are incredibly weak.[1] The reason is simple: the force of gravity is astonishingly feeble – 10,000 billion billion billion billion times feebler than the electric force that glues together the atoms in the human body.[2] 'You think Earth's gravity is really something when you're climbing the stairs,' says physicist Rainer Weiss of MIT. 'But as far as physics goes, it is a pipsqueak, infinitesimal, tiny little effect.'

An equivalent way of saying this is that space–time is incredibly stiff – a billion billion billion times stiffer than steel. It is easy to vibrate a drum skin, but incredibly hard to vibrate the drum skin of space–time. Although waving your hand in the air creates ripples in the fabric of space–time, only the most violent movements of mass imaginable, such as the merger of black holes, create gravitational waves powerful enough to be detected by twenty-first-century technology.

Black hole mergers, however, are likely to be extremely rare, which means that any whose gravitational waves are arriving at Earth are likely to have occurred immensely far away across the universe. Having spread through a mind-cringingly large volume of space, they would be diluted to infinitesimally small ripples. Everyone therefore expected that the first signal to be picked up by LIGO–Virgo would be barely perceptible amid the background 'noise'. But the signal Drago was staring at on his computer screen was far from weak and feeble. It was powerful and unmistakable. It virtually jumped out of the screen at him. So there was little doubt in his mind. It had to be a false alarm.

In search of a second opinion, Drago walked along the corridor to the office of a Swedish postdoctoral colleague. Andrew Lundgren pulled up the email on his computer and clicked on the links. When the pair of identical waveforms from Hanford and Livingston popped up on his screen, he was as unconvinced as Drago. The two men were of one mind: it was a fake.

There were two logical possibilities. Occasionally, in order to test the response of their instrument, the LIGO–Virgo engineers would inject a fake signal into the system. But it was their practice to flag up such a 'scheduled injection', and Drago and Lundgren could find no mention of such an event.

The second kind of fake signal was designed to ensure that the physicists would be able to distinguish a bone fide gravitational wave from a spurious artefact. A small team of physicists was responsible for such 'blind injections'. Sworn to secrecy, it would confess to creating a signal only if it was caught out or if a scientific paper claiming a detection was on the verge of appearing in a journal. On 16 September 2010, a blind injection had led to such a paper, which was about to be submitted to *Physical Review Letters* when it was revealed as a fake.

After an hour of discussion, Drago and Lundgren concluded that a blind injection was the most likely explanation for the signals they were looking at. The only thing to do was to phone Hanford and Livingston and check that everything was operating properly. It was 3.30am at Hanford, and Drago was only able to get a response from the control room at Livingston, where it was 5.30am. But it was quite enough. William Parker, the technician on duty, assured him that there were no problems with the instrument and confirmed that there had been no scheduled injection.

Drago and Lundgren decided to email the entire LIGO–Virgo collaboration about the alert and see what other people thought about it. 'Hi all,' typed Drago. 'Very interesting event in the last hour. Someone can confirm this is not a hardware injection?'

The rest of the day went by in a blur. In the US, people began to wake up and started to discuss the signal. There were so many emails for Drago to deal with that it was impossible to for him to do any other work. He was too busy to even feel excited, but he was still convinced that the signal was a fake.

Everything changed two days later, when a piece of heart-stopping news came through from the team of physicists whose job was to keep the scientists on their toes. There had been no blind injection.

The Albert Einstein Institute, not surprisingly, is adorned with more than one image of its famous namesake. That evening, as Drago left the building to go home, he passed a metre-high portrait of Einstein on the corridor wall. It seemed as if the great man, who had predicted the existence of gravitational waves almost exactly a century earlier, was smiling down at him mischievously and saying, 'I told you so.'

Princeton, New Jersey, July 1936

For ten minutes, Einstein's new Polish assistant, seated on the other side of his desk and partially eclipsed by teetering stacks of papers, had been singing Howard Percy Robertson's praises. Leopold Infeld appeared to have struck up quite a friendship with the man on his recent return from a sabbatical at Caltech. In fact, Einstein had that day watched the pair on the lawn from his office window. They had been in animated conversation in the bright July sunshine, Robertson puffing at regular intervals on his pipe before departing, his briefcase swinging, down the sweeping driveway of the Institute for Advanced Study.

Einstein knew who Howard Percy Robertson was. The young professor at Princeton University had impressed him the previous year when he had taken Einstein's own theory of gravity, made some plausible assumptions about the uniformity of matter throughout the universe and managed to obtain a 'cosmological solution' that perfectly explained Edwin Hubble's 1929 discovery of the expanding universe.

Infeld, leaning one way and then the other in a vain attempt to find a clear line of sight to his boss on the other side of the desk, said he had been telling Robertson about the paper Einstein had written on gravitational waves with his former assistant, Nathan Rosen. Robertson had spotted a flaw, which Infeld proceeded to describe nervously, not knowing how Einstein would respond. He need not have worried. 'Ah,' said Einstein. 'I too have recently become aware of that error.'

As soon as Infeld left his office, Einstein located the manu-script of the paper he had written with Rosen and began to revise it. The subject of gravitational waves had turned out to be far more problematical than he had expected. In 1916, within only a few months of presenting his revolutionary theory of gravity, the general theory of relativity, it had seemed blatantly obvious to him that it was possible to jiggle space–time and in doing so create ripples that propagated outwards at the speed of light. Such gravitational waves were completely analogous to the waves that rippled through the electromagnetic field and which, in 1862, Maxwell had triumphantly recognised as light.

However, Einstein had found it impossible to extract exact predictions from his full theory of gravity. This was not too surprising; in order to describe the gravitational field of any distribution of energy, he had been forced to replace Newton's one equation by ten. And if that was not enough of a complica-tion, all forms of energy – not just mass-energy – have gravity, including gravitational energy itself. Gravity creates gravity! To avoid his theory's calculation-defying 'non-linearity', Einstein had no choice but to consider only the case where gravity was weak and responsible for creating essentially no extra gravity. Not only did it cut down his ten equations to a manageable one, but it yielded a 'wave equation', just like the one found by Maxwell. Gravitational waves, it seemed, must exist.

The truth was that Einstein could be sure only that gravita-tional waves existed in this special instance of his theory when gravity was weak, but that would turn out to be no more than a mirage if they did not exist in the general case when gravity was strong. Proving that they were a feature of his full the-ory was a nightmare; the theory was so complex that all his intuition went out of the window. Nevertheless, the need to prove definitively that gravitational waves must exist had been nagging at the back of his mind for two decades. When he fled Nazi Germany and ended up at Princeton's Institute for Advanced Study following a sojourn in Southern California, it had come to the forefront of his thoughts once more.

Remarkably, he and Rosen – his first American student – had found a gravitational wave 'solution' to the full equations. It should have been a triumph, but unfortunately it contained a 'singularity', a place where the description of the wave blew up to infinity and so made no sense. Having set out to prove that gravitational waves exist, it appeared that he and Rosen had inadvertently proved that they did not exist after all.

Einstein was far from dismayed. Given that every other known 'field' – the electromagnetic field, air, water, and so on – could be rippled by waves, he viewed it as a remarkable finding that the gravitational field could not.* He had written to his friend, the quantum pioneer Max Born, back in Germany: 'Together with a young collaborator, I have arrived at an interesting result that gravitational waves do not exist.' He and Rosen had then co-written a paper, under the title 'Do Gravitational Waves Exist?', which they had submitted to the American journal *Physical Review* at the end of May 1936.

Einstein had been taken aback when, on 23 July, the editor of the journal had returned the manuscript. John Tate's covering letter said that an anonymous 'referee' who had been

---

* A 'field' is simply something that has a value at every point in space. That may be a number, as in the case of air, where each point possesses a number of a certain magnitude to represent the pressure. Or it could be a 'vector', as in the case of a magnetic field, where each point possesses a number representing the magnitude of the force and an arrow representing its direction. Think of it as a field of arrows.

sent the paper had pointed out that it contained an error. The referee claimed that the troubling infinity, which Einstein and Rosen had taken as proof that gravitational waves did not exist, was no more than a mathematical artefact that could easily be removed. Tate said he would be glad to have Einstein's reaction to the referee's comments and criticisms.

Rosen, who had left Princeton to take up an academic post in the Soviet Union, was spared the sight of his supervisor blowing his top. Back in Germany, journals such as *Annelen der Physik* published everything Einstein sent them, without question. If he had slipped up in a calculation, someone would simply publish a follow-up paper pointing out the error. This was the German way of doing science. Einstein could not believe the effrontery of *Physical Review* in rejecting the paper. He was not a prima donna. It was not a matter of pride. It was just not the way things should be done.

Einstein wrote a stinging letter back to Tate:

27 July 1936

Dear Sir,

We (Mr Rosen and I) had sent you our manuscript for publication and had not authorised you to show it to specialists before it is printed. I see no reason to address the – in any case erroneous – comments of your anonymous expert. On the basis of this incident I prefer to publish the paper elsewhere.

Respectfully,
Einstein

P.S. Mr Rosen, who has left for the Soviet Union, has authorised me to represent him in this matter.

Einstein was true to his word. Stung by his first encounter with 'peer review', he decided to send the paper to a small publication he had used before: the *Journal of the Franklin Institute*. Fortunately, he had not yet done so when Infeld relayed to him his conversation with Robertson; if he had, he would not have had the chance to correct his error.

The whole point of his equations of the gravitational field was that they were universal, or 'covariant'. Everyone in the universe must see the same physics and it should not depend on anyone's vantage point, technically known as a 'co-ordinate system'.* But the lucky flip side of this was that, if a calculation proved difficult or impossible in one co-ordinate system, it was necessary only to switch to another co-ordinate system in which the calculation might be easier to carry out. That was precisely what Robertson had suggested to Infeld that Einstein do.

By coincidence, the anonymous referee at *Physical Review* had suggested the same way out. At the time, Einstein had been too blinded by rage to take any notice, but when Infeld relayed his conversation with Robertson, he was happy to own up to his error.

He sat for an hour in his office, scribbling corrections in the margin of the paper. When he had finished, he leant back in his chair, lost in thought. So now he was saying that gravitational waves existed. He crossed out the title 'Do Gravitational Waves Exist?' and replaced it with 'On Gravitational Waves'.

He contemplated the intellectual journey that had led him to this point. In Berlin, in 1916, he had been sure that gravitational

---

* The location of a town might be specified as thirty kilometres north of London and thirty kilometres east of London, or it could be specified as 42.4 kilometres in a northeast direction. Both are examples of 'co-ordinate systems'.

waves existed, but the fact that he was not completely certain had been nagging in the back of his mind for two decades. This year, 1936, he and Rosen had convinced themselves that gravitational waves did not exist after all, and now, finally, he was sure that they did. However, the only plausible astronomical source he could imagine in 1918 – two stars locked in mutual embrace and spiralling together – would create gravitational waves of such mind-boggling weakness as to be impossible to detect in practice.*

He took his paper to Helen Dukas, his long-standing secretary, who had come with him from Germany. While she was retyping it, he returned to his office and wrote a covering letter to the editor of the *Journal of the Franklin Institute*.[3] In the early evening, as he left the Institute, he dropped it off with Dukas.

On Olden Lane, he passed a yellow bulldozer that was parked in the rutted earth in front of a house that was under construction. He squinted at it for a moment in the early-evening sunshine, wondering what it would be like to drive such a machine. He resolved that next time, if the workman was around, he would ask whether he could have a go. Fame, despite its quite considerable drawbacks, also had its perks.[4]

As he turned right into Mercer Street, heading for his house at number 112, he saw a figure come out of Marquand Park on the left, cross the road ahead and disappear down Springdale Road. It was the unmistakable figure of a pipe-smoking

---

* The irony is that Einstein did not believe in another prediction of his theory of gravity: black holes. Being the most compact possible objects, they can whirl around each other in much closer proximity than stars, creating much greater distortions of space–time and radiating far more intense gravitational waves.

Howard Robertson. How peculiar, he mused, that the man's suggestion had been so similar to that of the anonymous referee.[5]

This thought occupied his mind only briefly; the important thing was not how he had come to correct his paper with Rosen but that he had corrected it. Gravitational waves, he was now certain, must exist.

Hanover, September 2015

The weeks and months after 14 September 2015 were hard for Drago and everyone else on the LIGO–Virgo team. There was a long checklist of equipment and software to work through, and each item had to be examined to confirm that it had not been malfunctioning. 'Our first priority was making sure we weren't fooling ourselves,' said Keith Riles, a physicist at the University of Michigan and member of the LIGO Detection Committee.

There were no shortcuts – it was painstaking work that took a lot of time and effort. Could something on Earth have mimicked a cosmic signal? It was necessary to check seismic records across the world, to rule out the possibility that what had been detected was actually a small earthquake. Could something have happened simultaneously at each site, at the precise time the signal was registered? Had someone ridden past on a bike? Had a car hit a bump on a nearby road? In order to get identical signals at the two detectors, something identical would have to have happened at both Hanford and Livingston, which was stretching things a little. Nevertheless, extraordinary claims require extraordinary levels of confidence. All kinds of logs had to be consulted, microphone recordings listened to and CCTV video watched, in order to rule out the possibility of something like two doors being slammed at the same time.

The conclusion at the end of all this work was that, on average, random noise would generate two simultaneous and identical signals like the ones seen at Hanford and Livingston less than once every 200,000 years. The reality of the signal appeared to be beyond any doubt, but there was one final possibility that caused many members of the collaboration to have sleepless nights: that the signal they were seeing was malicious.

Could somebody have hacked the computers at each LIGO site and injected the signals? Such a possibility was difficult to disprove, but the team came to the conclusion that for a hacker to break into the computers and leave a false signal and no trace whatsoever that they had done so would have required a tremendous knowledge of many complex systems. 'Mission Impossible would be easier,' says Drago.

By the end of 2015, LIGO itself bolstered the team's confidence that the 14 September signal was genuine. There had been two more trigger signals from the giant rulers at the two sites – one on 12 October and one on 26 December – and each was exactly what was predicted for the merger of black holes. Both signals were weaker and more difficult to discern than the first one. What was the chance that mundane terrestrial problems with the instrument would conspire to create not one but three signals? 'We became a lot more confident,' says Drago.

Secrecy was paramount, and nobody in the collaboration was permitted to tell their friends or families what LIGO had discovered. For Drago this did not prove too difficult. The urge to tell people about the discovery was enormous, but he was so busy that everything except the work at hand was pushed from his mind.

Drago and his colleagues were in the rare position of knowing something nobody else in the world knew, and that nobody in the history of the world had known. The British biographer Peter Ackroyd has speculated that Isaac Newton may have savoured such a feeling, which might explain why he did not tell anyone about his universal law of gravity and his laws of motion until twenty-five years after his discovery. But such thoughts were far from Drago's mind. He was too busy to feel anything but exhausted. And, anyhow, he was sure the public would not have the slightest interest in the discovery. How wrong he was.

Although everyone was sworn to secrecy, it is very hard to keep a lid on things when around one thousand people from sixteen countries are involved. The task had been made even harder because of the need to inform people outside of the team. The time delay between the arrival of the signals at Hanford and Livingston did not allow the location of the source in the sky to be deduced precisely, but it narrowed it down to a relatively narrow band across the heavens. Astronomers at major observatories around the world were notified and asked to scan the band with their telescopes to see if they could spot anything unusual in the sky on or around 14 September.

Inevitably, rumours of a major discovery circulated within the scientific community. On 25 September, Lawrence Krauss of Arizona State University in Tempe tweeted: 'Rumor of a gravitational wave detection at LIGO detector. Amazing if true. Will post details if it survives.' The tweet caused journalists to begin phoning people on the collaboration. 'It was a little upsetting for everyone,' says Drago.

Gabriela González, the Argentinian–American spokesperson for LIGO, was dismayed. As early as 16 September, the day after Drago first saw the signal, she and four colleagues had emailed the LIGO–Virgo collaboration. 'We want to remind everyone that we need to maintain strict confidentiality,' they said. Premature publicity was a big worry for González because it is essential to be sure of a scientific result before announcing it. Nobody wants to have to retract an incorrect claim at a later date and get egg on their face. When journalists enquired, they received the official reply: 'We take months to analyse and understand foreground and background in our data, so we cannot say anything at this point.'

In early 2016, the signal first seen by Drago on 12 September 2015 received a name: GW150914. In the outside world, excitement that something important had been discovered was steadily mounting. On 11 January, Krauss upset everyone further, this time by tweeting: 'My earlier rumor about LIGO has been confirmed by independent sources. Stay tuned! Gravitational waves may have been discovered!! Exciting.' Not surprisingly, many on the LIGO team saw this as a brazen attempt to steal its thunder.

Excitement was reaching a crescendo when, on 8 February, a press conference was called for Thursday 11 February. It was timed to coincide with the publication of the paper announcing the discovery in *Physical Review Letters*.[6]

The press conference was held at the National Press Club in Washington DC. It kicked off at 10.30am Eastern Standard Time, with a brief introduction and video about the project. Kip Thorne, a theorist from Caltech in Pasadena, was on stage, as was Rainer Weiss from MIT. Although over the decades more than a thousand people had been involved in the project, Weiss and Thorne were widely considered to be the founding fathers of LIGO.

David Reitze, deputy director of LIGO, stood up, while behind him a TV screen showed a simulated picture of two black holes. He surveyed the expectant audience, paused for effect, and then spoke. 'Ladies and gentlemen. We have detected gravitational waves. We did it!'

'Many of us on the project were thinking if we ever saw a gravitational wave, it'd be an itsy bitsy little tiny thing; we'd never see it,' said Weiss. 'This thing was so big that you didn't have to do much to see it. I keep telling people I'd love to be able to see Einstein's face right now!'

Public interest exceeded Drago's wildest expectations. The discovery of gravitational waves, almost exactly one hundred years after Einstein had predicted them, was a huge international story, and the public were right to be excited. Science had gained an entirely new sense. Imagine if you had been deaf since birth and suddenly, overnight, were able to hear. This is what it was like for physicists and astronomers. Throughout history, they had been able to 'see' the universe with their eyes and their telescopes; now, for the first time, they could 'hear' it. Gravitational waves are the 'voice of space'.

Scientific discoveries are often over-hyped by the media, but a good case can be made that the detection of gravitational waves on 14 September 2015 was the most significant development in astronomy since Galileo turned his new-fangled telescope towards the heavens in 1609.

Drago was the first person in history to see the signature of gravitational waves. Before the moment he opened that email, they had been travelling for 1.3 billion years across space, and no human being knew of their existence. 'I could so easily have gone to lunch,' says Drago. 'Someone else would have been the first to see the signal, not me.'

Drago happened to be in the right place at the right time, something he freely acknowledges. 'When Christopher Columbus arrived in the Americas, there was obviously one person who happened to be the first to spot land,' he says. 'But everyone knows that it took a lot of people – in the case of Columbus, the crew of an entire ship – to get to that point. As it was with the discovery of the Americas, so it was with the discovery of gravitational waves.'

o   o   o

The two black holes whose merger created the gravitational waves which buffeted Earth on 14 September 2015 were the relics from two massive stars that exploded as 'supernovae'. Paradoxically, when such a star explodes, its 'core' implodes – in fact, the implosion is believed to drive the explosion. As the core undergoes runaway shrinkage, its gravity intensifies until it is so strong that nothing, not even light, can escape, and a black hole is born.

By its very nature, a black hole is black and tiny. Consequently, nobody has ever seen one directly, although a global network of radio dishes known as the Event Horizon Telescope is close to obtaining the first-ever image of Sagittarius A*, the 4.3-million-solar-mass 'supermassive' black hole in the dark heart of the Milky Way.[7]

Evidence of the existence of black holes has necessarily been indirect: commonly, the observation of a star, or stars, whirling around an invisible celestial object at unfeasibly high speed. However, the gravitational waves picked up on 14 September 2015 changed everything. Since their signature was precisely that predicted by Einstein's theory of gravity for merging black holes, it proved beyond doubt that black holes really do exist.[8]

The irony here is that black holes – a prediction of general relativity that Einstein did not believe – were confirmed by gravitational waves – a prediction that he did believe (or that he believed, did not believe and then believed again!).

The gravitational waves intercepted by LIGO on 14 September 2015 came from two black holes that were extraordinarily massive – twenty-nine and thirty-six times the mass of the Sun respectively. In the explosion of a supernova, most of the matter is blown into space and only a relatively small amount ends up sucked into a black hole. In fact, according to astrophysicists' estimates, the precursor stars must have been at least 300 times the mass of the Sun. Such stars are so rare as to be pretty much non-existent. However, there are strong theoretical reasons to believe that the first generation of stars that formed after the Big Bang – our Sun is a third-generation star spawned from the debris of two earlier generations – were far bigger than today's suns.[9] If LIGO's black holes really are relics of the very first stars, it would be like walking down London's Oxford Street and spotting two Roman legionaries among the crowds of shoppers who have miraculously survived in the city since the day the Empire's soldiers departed in AD 410.

It is perfectly possible that the pair of black holes had been orbiting each other for billions of years, all the while radiating gravitational waves, which sapped them of orbital energy, causing them to gradually spiral together. But it was only during the final ten or so orbits, each lasting about a hundredth of a second, that the convulsions of space–time were violent enough to create gravitational waves strong enough to be detectable on Earth.

Although the tenth-of-a-second-long pulse had been travelling for 1.3 billion years, it would have been missed had Advanced LIGO been turned on a month later. It seems that the collaboration was extraordinarily lucky. But physicists do not believe in luck. The fact that LIGO caught its quarry so soon after the instrument was turned on can mean only that black hole mergers are common. And this has proved to be the case.

Since the first black hole merger, eight more have been detected. Most significant was the detection on 17 August 2017 of a much weaker and longer-duration pulse of gravitational waves from a merger of two 'neutron stars'. Such stars, also spawned by supernova explosions, are formed when the core of a star is not massive enough for its gravity to crush it all the way down to a black hole. Typically, a neutron star is about the size of Mount Everest and so dense that a sugar-cube-sized volume of its material would weigh as much as the entire human race.

The crucial difference between a neutron star and a black hole is that it is not simply a bottomless pit in space–time but an object made of actual 'stuff'. Whereas a black hole merger produces nothing but gravitational waves – all matter in the vicinity having long ago been vacuumed up by each hole – a neutron star merger generates not only gravitational waves but a fireball of blisteringly hot matter. The radiation from the fireball was observed in the days after 17 August 2017 by about seventy ground-based and space-based telescopes, sensitive to different types of light. Most significant of all was the detection of an intense flash of high-energy gamma rays. This solved several cosmic mysteries in one go.[10]

In the late 1960s, the Americans launched satellites into orbit to detect the gamma rays from clandestine Soviet nuclear tests. To their shock and surprise, they discovered that around once a day there was a flash of gamma rays coming not up from the ground but down from space. The discovery of 'gamma-ray bursters' was not revealed to the astronomical community until the 1980s, and it was not until the 1990s that it became clear that they were at immense distances. It was then suggested that the most common type might be the result of a merger between two neutron stars. This theory was dramatically confirmed with the discovery of a gamma ray burst from the gravitational wave source detected on 17 August 2017.

But the gamma rays also revealed something else. The cores, or 'nuclei', of different atoms produce gamma rays at discrete energies, which provide a unique fingerprint for each kind of atom. And what gamma-ray astronomers saw was the sudden appearance of the fingerprint of gold; forged in the fireball was a mass of gold equal to about twenty times the mass of the Earth.

Astronomers have long known that all the elements heavier than hydrogen and helium were forged in the nuclear furnaces of stars, which, when they exploded, spewed them into space to be incorporated into successive generations of stars. But although nuclear astrophysicists had successfully identified the origin of pretty much all the ninety-two naturally occurring elements, they did not know the origin of gold. Now, at last, they do. Can there be a more striking connection between the mundane and close to home and the cosmic and far away? If you have a gold ring or necklace, its atoms were forged long before the Earth was born, in a fireball caused by the cataclysmic coalescence of two neutron stars.

The LIGO–Virgo team had predicted that 2017 would be the year its detector would reach the sensitivity to pick up a signal from the first neutron star merger, and it was proved to be correct. However, what the team failed to realise was that there existed a class of black hole mergers out there that was much more powerful and which they would easily spot with a less sensitive detector as early as 2015.

All the discoveries had been made with an extraordinary instrument. Although each LIGO site boasted a four-kilometre ruler made of a laser of light whose stretching and squeezing revealed a passing gravitational wave, in fact each site possessed two identical rulers, arranged in the shape of an 'L'. These were the arms of an 'interferometer', so called because it exploits the phenomenon of 'interference' to measure tiny changes in the paths taken by light.

When two waves – which could be waves of light, water or anything else – overlap, and the peaks of one coincide with the peaks of the second, they boost each other, in a process known as constructive interference. When the peaks of the first wave coincide with the troughs of the second, they cancel each other out, in a process known as destructive interference.

At each LIGO site, laser light is split in two, and half is sent down the evacuated tube of one arm of the interferometer and the other half down the other one. At the end of each arm, a suspended mirror reflects the light back the way it has come. The two halves are then combined and the brightness of the light measured. The key thing is that if one of the arms has been stretched relative to the other, the two light waves will not exactly match. If the peaks of one wave coincide with the troughs of the other, they will cancel each other out and the brightness measured will be zero. In fact, if the two light waves are even slightly out of step with each other, they will create an obvious change in the brightness of the light on recombination. In this way, it is possible to discern changes in the length of one arm relative to the other of a mere fraction of the 'wavelength' of the laser light – that is, a fraction of a thousandth of a millimetre.

Although measuring such a small change in the length of a four-kilometre-long arm may seem impressive, the detection of the gravitational waves on 14 September 2015 required measuring a hugely smaller change. Each arm was alternately stretched and squeezed not by a fraction of a thousandth of a millimetre but by a hundred-millionth of the diameter of a single atom. When you realise that it takes ten million atoms laid end to end to span the full stop at the end of this sentence, you will begin to appreciate the astonishing achievement of detecting gravitational waves.* 'The signals are infinitesimal. The sources are astronomical. The sensitivities are infinitesimal. The rewards are astronomical,' writes Janna Levin in *Black Hole Blues*.[11]

Given the scale of the achievement, it was no accident that the 2017 Nobel Prize in Physics was awarded for the discovery of gravitational waves. The three founding fathers of LIGO were considered to be Weiss, Thorne and a Scottish experimental physicist. Ronald Drever had Alzheimer's disease and was in a care home near Glasgow.[12] Sadly, he died only months before the awarding of the Nobel Prize, and Weiss and Thorne instead shared the prize with Barry Barish.

The aim now is not only to improve the sensitivity of the LIGO detectors but to bring online more detectors around the world, to better locate the source of any burst of gravitational waves. The European Virgo instrument became operational soon after Advanced LIGO and participated in some

---

* Einstein predicted the laser. How astonishing, then, that a prediction of Einstein's (gravitational waves) that confirmed another prediction of Einstein's (black holes) was confirmed by an instrument that used yet another prediction of Einstein's (the laser)!

of its discoveries, such as that of the neutron star merger. The Kamioka Gravitational Wave Detector (KAGRA) in Japan is set to join the network in 2020, and one in India will join by 2025.

The merger of neutron stars and black holes did not surprise the LIGO and Virgo experimenters, who had hoped to detect them. But what is most exciting about the gravitational wave 'window' they have opened up on the universe is the possibility of seeing things nobody expects and of being utterly surprised. 'The discoveries, made by our thousand-strong international team, are just the start,' says Sheila Rowan, LIGO physicist and Scientific Advisor to the Scottish Government. 'There should be lots more amazing stories to come.'

We can learn a lesson from light. Once upon a time, we knew only about the light we could see with our eyes. Then we discovered that this is merely a tiny fraction of the 'electromagnetic spectrum' and that in addition to the colours of the rainbow, there are a million other 'invisible colours'. When we learnt to look at the universe with artificial eyes that could see such colours – gamma rays, X-rays, ultraviolet, infrared, radio waves, and so on – we discovered all kinds of unexpected things. We discovered gamma-ray bursters and 'pulsars'. We discovered 'quasars' and supermassive black holes. We discovered the relic 'afterglow' of the Big Bang fireball and planets around other stars – more than four thousand of them at the last count.

Now, with the success of LIGO and Virgo, we stand at the dawn of a new era in astronomy. 'We know about black holes and neutron stars, but we hope there are other phenomena we can see because of the gravitational waves they emit,' says Weiss.

It is as if we have been deaf and have now gained a sense of hearing, but at present our hearing is crude and rudimentary. At the very edge of audibility, we have heard a sound like a

distant rumble of thunder, but we have yet to hear the equivalent of birdsong, of a baby crying or of a piece of music. As
LIGO and Virgo and other gravitational wave experiments
around the world increase their sensitivity, who knows what
wonderful things we will discover as we tune into the cosmic
symphony?

10

The poetry of logical ideas

The miracle of the appropriateness of the language
of mathematics for the formulation of the laws
of physics is a wonderful gift which we neither
understand nor deserve.

EUGENE WIGNER

Pure mathematics is, in its way, the
poetry of logical ideas.

ALBERT EINSTEIN

The stories I have recounted have, I hope, illustrated the central magic of science – its ability to predict the existence of things that, when people go and look for them, turn out to exist in the real world. It is an ability that is so magical that even the exponents of science can scarcely believe it. As mentioned before, the question of why it is so hard for physicists to believe the predictions of their own theories was pondered by Steven Weinberg in his book, *The First Three Minutes*.[1] What puzzled him about the birth of the universe was why the prediction of the afterglow of the Big Bang in 1948 had been ignored and the cosmic background radiation stumbled upon by accident only in 1965. 'The problem is not that we take our theories too seriously,' he concluded, 'but that we do not take them seriously enough.'

It is easy to see why physicists find it so hard to believe their theories. After all, how is it possible that arcane mathematical formulae scrawled across blackboards or whiteboards can have anything whatsoever to do with real things in the real world?

How can it be that the universe out there has a mathematical twin down here, which mimics it in every conceivable way?

This remarkable fact was apparent even to Galileo in the seventeenth century: 'Philosophy is written in the grand book – I mean the universe – which stands continually open to our gaze, but it cannot be understood unless one first learns to comprehend the language and interpret the characters in which it is written. It is written in the language of mathematics, and its characters are triangles, circles and other geometric figures, without which it is humanly impossible to understand a single word of it; without these, one is wandering about in the dark labyrinth.'[2]

Isaac Newton, who was born in the year Galileo died, was successful in expressing the laws of motion and of gravity as precise mathematical formulae. And in the years that followed, mathematics scored more and more successes in describing ever greater tracts of physical reality. In the nineteenth century, this success was exemplified by the triumph of Maxwell's equations of electromagnetism, and in the twentieth century by the equations of quantum theory and Einstein's theory of gravity – the general theory of relativity.

Undoubtedly, however, the most powerful demonstration of the deep connection between mathematics and the physical universe was the Dirac equation. Despite the fact that his triumphant description of the electron, compatible with both quantum theory and Einstein's special theory of relativity, was conjured out of thin air purely for mathematical consistency, it nevertheless predicted the existence not only of quantum spin but of a hitherto unsuspected universe of antimatter. Dirac, as amazed as everyone else, concluded, 'God used beautiful mathematics in creating the world.'[3]

In the 1930s, Eugene Wigner wrote a famous essay entitled 'The Unreasonable Effectiveness of Mathematics in the Natural Sciences'. 'The enormous usefulness of mathematics in the natural sciences is something bordering on the mysterious,' he wrote. 'And there is no rational explanation for it.'[4]

Einstein echoed Wigner's observation. 'How can it be', he asked, 'that mathematics, being after all a product of human thought which is independent of experience, is so admirably appropriate to the objects of reality?'[5] Famously, he also remarked that 'The most incomprehensible thing about the world is that it is comprehensible.' And by 'comprehensible', he implicitly meant comprehensible by mathematics.

'Our work is a delightful game,' said Murray Gell-Mann, who won the Nobel Prize for proposing the existence of 'quarks', the ultimate building blocks of matter. 'I am frequently astonished that it so often results in correct predictions of experimental results. How can it be that writing down a few simple and elegant formulae, like short poems governed by strict rules such as those of the sonnet or the waka, can predict universal regularities of Nature?'[6]

So why is mathematics so unreasonably effective in the natural sciences? Why is the universe mathematical? The first thing to say is that not everyone believes these are valid questions. According to Stephen Wolfram, the billionaire creator of the symbolic computer language Mathematica, the universe is not mathematical; it simply looks that way.

Wolfram points out that most of what is happening in the universe, such as the turbulence in the atmosphere and biology, is far too complex to be encapsulated by mathematical physics. Most physicists would argue that this is because, at present, we lack mathematical tools of sufficient sophistication, but that this is only a temporary situation and that one day we will obtain such tools. Wolfram begs to differ. He thinks the reason we cannot describe complex phenomena like turbulence and biology with mathematics is because it is impossible.

According to Wolfram, we are in the position of a drunk man hunting for his dropped car keys on a street at midnight. He looks in the pool of light under a street light, for no other reason than that is the only place he can reasonably look. Similarly, claims Wolfram, we use mathematics to describe the only part of the universe that it is describable by mathematics.

Bertrand Russell, the twentieth-century British mathematician and philosopher, would have agreed with Wolfram. 'Physics is mathematical', he said, 'not because we know so much about the physical world, but because we know so little. It is only its mathematical properties that we can discover.' Percy Bridgman, an American physicist, said something similar: 'It is the merest truism, evident at once to unsophisticated observation, that mathematics is a human invention.'[7] Arthur Eddington put it this way: 'The mathematics is not there till we put it there.'

If, as Wolfram claims, the universe is not fundamentally mathematical, it is without doubt doing something far from random. There is a regularity. There are rules more basic than mathematical equations. They are encapsulated in simple computer programs, and it is such programs that Wolfram thinks are orchestrating everything we see in the universe. These programs are 'recursive', which means their output is continually fed back in as their input, like a snake eating its own tail. In the early 1980s, Wolfram had played around with such simple programs on one of the first personal computers and discovered that, occasionally, they can generate infinite complexity and novelty. It was such a striking finding that it had caused him to wonder whether this might be the secret of how nature creates a rose, a newborn baby or a galaxy.

Generally, the only way to discover the consequences of such a program is to run it and find out. According to Wolfram, this is true of most of what is going on in the universe, which he terms 'computationally irreducible'. However, for a small subset of programs, it is possible to discover their outcomes in advance of running them. Wolfram terms these 'computationally reducible'. The special shortcut that enables the prediction of their outcomes is none other than mathematical physics.

Most physicists disagree with Wolfram and believe that the universe is indeed mathematical. So the question remains: Why is mathematics so effective in the natural sciences? Why does the central magic of science work? There have been many attempts over the years to answer these questions. A remarkable one has come from the Swedish–American physicist Max Tegmark, and it involves multiple universes.

In recent years, evidence has been mounting from many different directions that our universe is not the only one. For instance, the fact that the universe was born 13.82 billion years ago means that we can see only those galaxies whose light has taken less than 13.82 billion years to reach the Earth. The 'observable' universe is therefore bounded by a 'horizon' rather like the membrane of a soap bubble; beyond the horizon are galaxies whose light has not yet got here. In other words, there are other domains – possibly an infinite number of them – like our observable universe, but with different stars and galaxies. The ensemble of such universes is called the 'multiverse'.

In addition to this rather trivial multiverse, physicists have reason to believe there may be other universes with different numbers of dimensions, different laws of physics, and so on. No one knows yet how all these multiverse ideas fit together. It is an emerging 'paradigm'.

Tegmark takes this idea to its logical conclusion and suggests that there may be an 'ultimate ensemble' of universes, in which every piece of mathematics is actualised. So, for instance, there is a universe that contains only flat space, or 'Euclidean', geometry, another that contains only arithmetic, and so on. Nothing happens in most of these universes because the rules are too simple to create anything of interest. Only in universes with mathematics as complex as the 'theory of everything' that is believed to orchestrate our universe is it possible for the emergence of interesting things, like stars and planets and life. The reason we find ourselves in such a universe, according to the topsy-turvy logic of the 'anthropic principle', is that in any simpler universe we would not have arisen to notice the fact.

Tegmark is claiming that mathematics is unreasonably effective in the physical sciences for a completely trivial reason: because mathematics *is* physics. 'Our successful theories aren't mathematics approximating physics, but mathematics approximating mathematics,' he says.[8] Heinrich Hertz had a similar thought a century and a half earlier: 'One cannot escape the feeling that these mathematical formulas have an independent existence and an intelligence of their own, that they are wiser than we are, wiser even than their discoverers.'

Many physicists would argue that a multiverse containing a possible infinity of universes, most of which are devoid of anything interesting, is a high price to pay for an explanation of Wigner's remark, but many would accept that, for some mysterious reason, the universe appears to be a manifestation of an underlying mathematical structure. Powerful evidence that this is the case is outlined by science writer Graham Farmelo in his book, *The Universe Speaks in Numbers*. Not only does mathematics provide insights into physics, says Farmelo, but physics provides insights into mathematics; it is a two-way street. The most striking example of this is 'string theory', which views the fundamental building blocks of matter not as point-like particles but as strings of mass-energy vibrating in ten-dimensional space–time. Although the theory is yet to contribute any testable predictions for physics, it has already opened up whole new vistas of enquiry for pure mathematics.

Tegmark and Wolfram's explanations for the unreasonable effectiveness of mathematics are not the only ones. The late American physicist Victor Stenger was fond of pointing out that the physics we have discovered is actually no more than the physics of 'nothing'.

Recall that in 1918, Emmy Noether showed that the great conservation laws of physics are merely a consequence of deep symmetries. So, for instance, the law of conservation of energy, which states that energy can neither be created nor destroyed, is a consequence of time-translation symmetry – the fact that the outcome of an experiment does not depend on when it is carried out. The conservation of momentum is the consequence of space-translation symmetry – the fact that an experimental result does not depend on its location in space, whether the experiment is done in London or in New York. 'If mathematics is the language of nature, symmetry is its syntax,' says Gian Francesco Giudice.[9] What is striking about these symmetries, Stenger pointed out, is that they are also the symmetries of an entirely empty universe. In a featureless void, after all, every time is exactly like every other time and every location is precisely like every other location.

In addition to such global symmetries, our universe maintains local symmetries. And as explained earlier, the fundamental forces of nature exist merely to ensure that such local 'gauge invariance' is enforced everywhere in space and time.[10] These local symmetries, like the universe's global symmetries, are also the symmetries of empty space.

There are, of course, other fundamental laws of nature besides those that are merely the consequence of deep symmetries. However, the British chemist Peter Atkins points out that these, too, also arise from nothing – or at least, they are not the substantive prescriptions that at first sight they appear to be.

Take the law which dictates the path taken by a ray of light through a medium such as glass, better known as the law of refraction. It turns out that there is another, entirely equivalent way of determining the trajectory taken by light: it follows the path that takes the least time. A little thought reveals that the only way a light beam can possibly do this is by trying all possible routes through, say, a piece of glass, to determine the quickest path. It may seem mad, but this is pretty much what light does.

The critical thing you need to know is that light is a wave. Now, imagine that the light travelling through a block of glass takes all possible paths between points A and B. The wavelength of light is small, which means that light rays following neighbouring paths differ substantially in the location of their peaks and troughs. In fact, for each path there is a neighbouring one for which the peaks of one wave coincide with the troughs of the other, and vice versa; consequently, they cancel each other out. The only path that does not suffer such 'destructive interference' is the path of shortest time.

In a sense, the real law of refraction is that there is no law; the light follows every possible route and the passive phenomenon of interference culls all but the shortest-time path. According to Atkins, nature's law of refraction is nothing but a law of laziness, or 'indolence'.

The path followed by a ray of light may not seem to have any wider significance, but it does. The reason is that in quantum theory – our very best description of the microscopic world of atoms and their constituents – the behaviour of the building blocks of matter is described by a wave function. According to the Schrödinger equation, this spreads through space, and where the wave is big, or has a large amplitude, there is a high probability of finding a particle such as an electron, and where it is small there is a low chance.

In this 'many histories' interpretation of quantum theory, which was devised by Richard Feynman, a particle travelling between points A and B tries every conceivable path. And just as in the light example, one trajectory is picked out by interference. Rather than being the one that takes the least time, it is the one that takes the least 'action', but the principle is the same.* And just as in the example of the light ray, nature's law is nothing but a law of indolence.

So not only are many of the laws of physics the same as the laws of nothing, as Stenger maintained; those that remain are also nothing – or born of indolence – as Atkins claims. 'Nothing is extraordinarily fruitful,' says Atkins. 'Within the infinite compass of nothing lies potentially everything, but it is an everything lurking wholly unrealised.'[11]

Even though we live in a universe whose laws are arguably the same as they would be if the universe was an empty void,

---

* In physics, action is a quantity that involves both the potential energy and the energy of motion of a particle.

there remains one sticky question: Why, rather than living in a universe of nothing, do we live in a universe of organised nothing? The answer, of course, is that nobody knows.

As for the central magic of science, what progress have we made in understanding it? Recall that Paul Murdin and Louise Webster discovered Cygnus X-1, the first black hole candidate in the Milky Way, in 1971. The existence of such an entity had been predicted in 1916 by Karl Schwarzschild, while he was on the Eastern Front, dying of an auto-immune disease that covered his skin in ugly, painful blisters. Murdin, like every other scientist who has ever confirmed a scientific prediction, expressed amazement at his discovery. 'The surprising thing is that black holes turn out to be real objects,' says Murdin. 'Incredibly, they actually exist!' The fact remains that physicists are just as gobsmacked as they were in Urban Le Verrier's day about the unreasonable effectiveness of mathematics and the incredible predictive power of science. The central magic remains as magical, and as inexplicable, as ever.

Notes

Introduction: The central magic of science

1 *A Shadow Passes* by Eden Phillpotts (cited in *The Strange Death of Fiona Griffiths* by Harry Bingham).
2 *The Ancestor's Tale: A Pilgrimage to the Dawn of Evolution* by Richard Dawkins (Weidenfeld & Nicolson, London, 2005).
3 *The Ascent of Gravity: The Quest to Understand the Force that Explains Everything* by Marcus Chown (Weidenfeld & Nicolson, London, 2017).
4 *Pensée d'un biologiste* by Jean Rostand (1939).

1: Map of the invisible world

1 *Philosophy of the Inductive Sciences, Volume 2* by William Whewell (1847, p. 62).
2 *Astronomy for Amateurs (1915)* by Camille Flammarion (Kessinger Publishing, Whitefish, 2008, p. 171).
3 *The Planet Neptune: An Exposition and History* by John Pringle Nichol (Kessinger Publishing, Whitefish, 2010, p. 90).
4 Venetia Burney talking at eighty-five about how, aged eleven, she came to name Pluto: https://www.nasa.gov/mp3/141071main_the_girl_who_named_pluto.mp3.
5 'Evidence for a Distant Giant Planet in the Solar System' by Konstantin Batygin and Mike Brown (*Astronomical Journal*, vol. 151, 20 January 2016, p. 22).

2: Voices in the sky

1 *The Feynman Lectures on Physics, Volume II* by Richard Feynman, Robert Leighton and Matthew Sands (Addison-Wesley, Boston, 1989).
2 'Tartan rosette: an animated recreation of James Clerk Maxwell's "Tartan Ribbon" photograph, the world's first colour photograph' by Ron Pethig and David Peacock (James Clerk Maxwell Foundation: https://vimeo.com/130333096).
3 Davy's invention of the safety lamp created a priority dispute, since the engineer George Stephenson came up with a similar design in the same year.

4 Ernest Rutherford had a similar experience. Always coming second in exams, he lost out on a coveted 1851 Great Exhibition scholarship to Britain to a fellow New Zealander. But at the eleventh hour, the other man, who had recently married, decided to take up a steady, well-paid government post in Auckland. Rutherford was digging potatoes on the family farm near Nelson when he got the good news. In later life, as he stood at the pinnacle of world science as Lord Rutherford, the greatest experimental physicist of the twentieth century, the thought of how nearly his life could have gone another way could bring him to tears.

5 In *Frankenstein: Or the New Prometheus*, Mary Shelley uses Humphry Davy as the model for the character of Professor Waldman. Victor Frankenstein studies with Waldman at the University of Ingolstadt in Bavaria. His lecture, which is very similar to Davy's 'A Discourse, Introductory to a Course of Lectures on Chemistry' in 1802, inspires Frankenstein to search for the secret of life.

6 Volta himself gave Faraday a battery when he and Davy met the great man, then aged sixty-nine, in Italy in June 1814.

7 'Faraday's Notebooks: Electromagnetic Rotations' (Royal Institution of Great Britain: http://www.rigb.org/docs/faraday_notebooks__rotations_0.pdf).

8 It was James Clerk Maxwell who called Ampère 'the Newton of Electricity', mainly because the Frenchman formulated a law which expressed the electrical force between 'current elements' in much the same way that Newton formulated a law that expressed the gravitational force between masses.

9 Letter to Christian Schönbein (13 November 1845), *The Letters of Faraday and Schoenbein, 1836–1862* (1899, p. 148).

10 The South Kensington Museum was renamed the Victoria and Albert Museum in 1899.

11 *What Mad Pursuit: A Personal View of Scientific Discovery* by Francis Crick (Basic Books, New York, 1990).

12 Now renamed 16 Palace Gardens Terrace.

13 Granulomatosis with polyangiitis (Wegener's granulomatosis): https://www.nhs.uk/conditions/granulomatosis-with-polyangiitis/.

14 *The Heinrich Hertz Wireless Experiments at Karlsruhe in the View of Modern Communication* by D. Cichon and W. Wiesbeck (University of Karlsruhe, October 1995).

15 *Dynamic Fields and Waves* by Andrew Norton (CRC Press, Boca Raton, 2000, p. 83).

3: Mirror, mirror on the wall

1 *The Physicist's Conception of Nature: Symposium on the Development of the Physicist's Conception of Nature in the Twentieth Century*, edited by Jagdish Mehra (Springer, 1973, p. 271).

2 'Paul Dirac: The Purest Soul in Physics' by Michael Berry (*Physics World*, 1 February 1998: https://physicsworld.com/a/paul-dirac-the-purest-soul-in-physics/).

3 Robert Andrews Millikan was a controversial scientist, and many accusations have been made against him since his death in 1953: that he was a misogynist, an anti-Semite and that he committed scientific fraud by discarding data that did not fit his hypothesis in the famous oil-drop experiment, in which he measured the charge on the electron. On the first two charges, Millikan was probably a man of his time, with the views to match. And the latter charge may not hold up ('In Defense of Robert Andrews Millikan' by David Goodstein (*Engineering and Science*, vol. 63 (4), p. 30; http://calteches.library.caltech.edu/4014/1/Millikan.pdf)). However, one thing Millikan is guilty of is taking full credit for measuring the charge on the electron. The experiment was carried out with a graduate student, Harvey Fletcher. Two things were discovered: the charge on the electron; and that an oil drop, suspended in mid-air by an electric force field balancing gravity, was buffeted by air molecules – an effect known as 'Brownian motion'. In order to use a scientific paper as a doctoral thesis, Fletcher needed to be the sole author. Millikan assigned Fletcher the Brownian result and himself the electric charge result, knowing that the latter was the most important. He was right, and while he achieved fame when he won the Nobel Prize, Fletcher was forgotten.

4 'Seth Neddermeyer (1907–88)' interviewed by John Greenberg (California Institute of Technology Archives, 7 May 1984: http://oralhistories.library.caltech.edu/199/1/neddermeyer_oho.pdf).

5 'Carl Anderson (1905–91)' interviewed by Harriett Lyle (California Institute of Technology Archives, 9 January–8 February 1979: http://oralhistories.library.caltech.edu/89/1/oh_anderson_c.pdf).

6 'Recollections of 1932–33' (*Engineering and Science*, vol. 46 (2), p. 15: http://calteches.library.caltech.edu/3353/1/Recollections.pdf).

7 Nowadays, we know that cosmic rays are high-energy atomic nuclei – mostly hydrogen nuclei – from space. The lower energy particles come from the Sun, and high-energy particles – some of which have energies tens of millions of times higher than anything achievable at accelerators such as the Large Hadron Collider – come from deep space. One extragalactic source, identified in 2018, is the supermassive-black-hole-powered 'blazar' galaxy

TXS 0506+056 ('Neutrino Emission from the Direction of the Blazar TXS 0506+056 Prior to the IceCube-170922A Alert' by the IceCube Collaboration (23 July 2018: https://arxiv.org/pdf/1807.08794.pdf)).

8 'Possible Existence of a Neutron' by James Chadwick (*Nature*, vol. 129, 27 February 1932, p. 312).

9 *It Must Be Beautiful: The Great Equations of Modern Science*, edited by Graham Farmelo (Granta Books, London, 2002).

10 *The Strangest Man: The Hidden Life of Paul Dirac, Quantum Genius* by Graham Farmelo (Faber & Faber, London, 2010).

11 Ibid.

12 An equivalent description of the quantum world, known as 'matrix mechanics', was devised by Werner Heisenberg, Max Born and Pascual Jordan, also in 1925.

13 *Antimatter* by Frank Close (Oxford University Press, Oxford, 2007).

14 The effect was first revealed in 1896 by the Dutch physicist Pieter Zeeman. Electrons in atoms are permitted by the laws of quantum theory to occupy only a discrete set of orbits, each with its own energy. When an electron makes a transition from one energy level to another, it either emits or absorbs a photon of energy equal to the difference in the two levels. However, Zeeman discovered that if a substance such as sodium is placed in a strong magnetic field, the energy of its characteristic photons is no longer well defined. The 'spectral lines' they produce in a 'spectroscope', rather than being sharp, are blurred. Later, more powerful instruments revealed that the lines are split into two, or more. This was a deep puzzle. As Wolfgang Pauli recorded, 'A colleague who met me strolling rather aimlessly in the beautiful streets of Copenhagen said to me in a friendly manner, "You look very unhappy", whereupon I answered fiercely, "How can one look happy when he is thinking about the anomalous Zeeman effect?"' The splitting of spectral lines, however, was exactly what would be expected if the electron is a tiny magnet that can be aligned in the direction of the magnetic field or against it, the two possibilities having a slightly different energy.

15 Or it may have been early December.

16 'The Quantum Theory of the Electron' by P. A. M. Dirac (*Proceedings of the Royal Society A*, vol. 177, issue 778, 1 February 1928: http://rspa.royalsocietypublishing.org/content/royprsa/117/778/610.full.pdf).

17 'Quantised Singularities in the Electromagnetic Field' by P. A. M. Dirac (*Proceedings of the Royal Society A*, vol. 133, issue 821, 1 September 1931).

18 'Lectures on Quantum Mechanics', Princeton University, October 1931.

19 'The Standard Model' by Sheldon Glashow (*Inference*, 2018: http://inference-review.com/article/the-standard-model).

20 *Life on the Mississippi* by Mark Twain (1883).

21 Actually, the tracks of positrons had been seen but not recognised as early as 6 May 1929. In Leningrad, Russian physicist Dimitry Skobeltzyn had been using a cloud chamber to investigate high-energy 'gamma rays', but they not only ejected electrons from atoms in the gas of his chamber but from the chamber walls. To get rid of the latter, which interfered with his measurements, he had the idea of using a magnetic field to sweep them away; it was then that he saw the inexplicable tracks of electrons curving the wrong way.

22 *It Must Be Beautiful*, edited by Graham Farmelo (Granta Books, London, 2002).

23 'The Reason for Antiparticles' by Richard Feynman (Dirac Memorial Lecture, University of Cambridge, 1986).

24 Interview with Paul Dirac by Thomas Kuhn at Dirac's home, Cambridge, 7 May 1963.

25 'Pretty Mathematics' by P. A. M. Dirac (*International Journal of Theoretical Physics*, vol. 21, issue 8–9, August 1982, p. 603).

26 *The Strangest Man: The Hidden Life of Paul Dirac, Mystic of the Atom* by Graham Farmelo (Faber & Faber, London, 2009, p. 435).

27 'The Evolution of the Physicist's Picture of Nature' by Paul Dirac (*Scientific American*, vol. 208, May 1963, p. 45).

28 From a conversation between Gerard 't Hooft and Dirac biographer Graham Farmelo, told to me by Farmelo.

4: Goldilocks universe

1 *Home Is Where the Wind Blows: Chapters from a Cosmologist's Life* by Fred Hoyle (University Science Books, California, 1994, p. 264).

2 It did not go unnoticed that Hoyle had pursued his astronomical interests while in the US on official radar business for the Admiralty. Back in the UK, he was asked to explain his visit to the Mount Wilson Observatory, having been reported by someone at the British Embassy in Washington DC. Thinking on his feet, Hoyle replied that he was interested in the well-known temperature inversion in the Los Angeles Basin, which caused a jump in the density of the air, with possible consequences for the propagation of radar pulses. Anomalous propagation of such radar pulses had been the subject of the conference he had attended in Washington DC; by neatly tying together his astronomical excursions and his 'tour of duty', he escaped any official reprimand and punishment.

3 In fact, very heavy elements such as californium, plutonium, einsteinium and fermium were found to have been created in the fall-out from the world's first large H-bomb test on Enewetak Atoll in the Pacific on 1 November 1952.

4 A key discovery that would be made by Paul Merrill in 1952 is the fingerprint of technetium in the light of stars. Since the element disintegrates, or decays, in only a few hundred thousand years, it can persist in stars only if it is made continually.

5 'The Chemical Composition of the Stars' by Fred Hoyle (*Monthly Notices of the Royal Astronomical Society*, vol. 106, 1946, p. 255).

6 Hoyle proved to be right even when he was wrong, as dense hydrogen gas does turn out to exist in space. Such 'Giant Molecular Clouds' are the nurseries where stars are born. However, the idea of such clouds proved so controversial for decades after Hoyle and Lyttleton came up with the idea that the only place Hoyle could present it was in a science-fiction novel, *The Black Cloud* (William Heinemann, London, 1957).

7 'Experimental and Theoretical Nuclear Astrophysics: The Quest for the Origin of the Elements' by William Fowler (Nobel Lecture, 8 December 1983: https://www.nobelprize.org/uploads/2018/06/fowler-lecture.pdf).

8 Ward Whaling interviewed by Shelley Irwin (California Institute of Technology Archives, April–May 1999: http://oralhistories.library.caltech.edu/122/1/Whaling_OHO.pdf).

9 *Home Is Where the Wind Blows: Chapters from a Cosmologist's Life* by Fred Hoyle (University Science Books, California, 1994, p. 265).

10 It actually took about three months to pin down the result precisely. Hoyle was back in Cambridge by the time Whaling and his team wrote their paper, to which they added Hoyle's name. 'A State in Carbon-12 Predicted from Astrophysical Evidence' by F. Hoyle, D. N. F. Dunbar, W. A. Wenzel and W. Whaling (*Physical Review*, vol. 92, no. 4, 1953, p. 1095a).

11 William Fowler interviewed by Charles Weiner (*American Institute of Physics*, 6 February 1973: https://www.aip.org/history-programs/niels-bohr-library/oral-histories/4608-4).

12 'When Is a Prediction Anthropic? Fred Hoyle and the 7.65MeV Carbon Resonance' by Helge Kragh (University of Aarhus, May 2010: http://philsci-archive.pitt.edu/5332/1/3alphaphil.pdf).

13 'Synthesis of the Elements in Stars' by E. M. Burbidge, G. R. Burbidge, W. A. Fowler and F. Hoyle (*Reviews of Modern Physics*, vol. 29, 1957, p. 547). Similar ideas were also published almost simultaneously by Alastair Cameron in 'Nuclear Reactions in Stars and Nucleogenesis' (*Proceedings of the Astronomical Society of the Pacific*, vol. 69, 1957, p. 201).

5: Ghost busters

1 *Proceedings of the 13th International Conference on Neutrino Physics &
Astrophysics*, edited by Jacob Schneps et al. (World Scientific, p. 575, 1989).

2 'Frederick Reines Dies at 80; Nobelist Discovered Neutrino' by John
Noble Wilford (*New York Times*, 28 August 1998: http://www.nytimes.
com/1998/08/28/us/frederick-reines-dies-at-80-nobelist-discovered-
neutrino.html).

3 KSU Physics Ernest Fox Nichols Lecture by Herald Kruse, 1 March 2010:
https://www.phys.ksu.edu/alumni/nichols/2010/kruse-lecture.pdf.

4 'Deadly Legacy: Savannah River Site Near Aiken One of the Most
Contaminated Places on Earth' by Doug Pardue (*The Post and Courier*,
21 May 2017: https://www.postandcourier.com/news/deadly-legacy-
savannah-river-site-near-aiken-one-of-the/article_d325f494-12ff-11e7-9579-
6b0721ccae53.html).

5 Ibid.

6 Quoted in *Great Physicists* by William Cropper (Oxford University Press,
New York, 2001, p. 257). Pauli is reputed to have said this during his student
days in Munich, when Einstein lectured at a crowded colloquium.

7 'The Average Energy of Disintegration of Radium E' by C. D. Ellis and W. A.
Wooster (*Proceedings of the Royal Society*, vol. A 117, 1 December 1927, p. 109).

8 Letter to Oskar Klein, 10 March 1930.

9 'The Romance that Led to a Legendary Science Burn' by Esther Inglis-Arkell
(*Gizmodo*, 17 February 2015: https://io9.gizmodo.com/the-romance-that-led-
to-a-legendary-science-burn-1686216120).

10 *No Time to Be Brief: A Scientific Biography of Wolfgang Pauli* by Charles Enz
(Oxford University Press, 2010, p. 210).

11 Letter from Wolfgang Pauli to Gregor Wentzel, 7 September 1931.

12 Pauli was so badly affected by the break-up of his marriage that after his
divorce on 26 November 1930, he appealed to none other than Carl Jung for
counselling. The great psychoanalyst recognised immediately that Pauli was an
academic whose social life had been neglected at the expense of his intellectual
life. He judged that Pauli needed a woman, not a man, to bring some balance
back into his affairs, and referred him to his pupil. Erna Rosenbaum, a
relatively inexperienced analyst, asked Pauli to send her his dreams so that she
might interpret them, something she may have regretted when, over time, he
sent her a total of 1,300 (*No Time to Be Brief: A Scientific Biography of Wolfgang
Pauli* by Charles Enz (Oxford University Press, 2010, p. 243)).

13 *The Historical Development of Quantum Theory, Volume 1, Part 1. The Quantum
Theory of Planck, Einstein, Bohr, and Sommerfeld. Its Foundation and the*

Rise of Its Difficulties, 1900–25 by Jagdish Mehra and Helmut Rechenberg (Springer, Heidelberg, 1982).

14 'Pauli's Letter of 4 December 1930: The Proposal of the Neutrino': http://www.physics.princeton.edu/~mcdonald/examples/EP/pauli_neutrino_30_english.pdf.

15 h/2*pi, where h is Planck's constant, a tiny quantity equal to 6.62607004 × 10^{-34} m^{2}kg/s.

16 *The Last Man Who Knew Everything* by David Schwartz (Basic Books, New York, 2018).

17 'Fundamental Forces' (*Eric Weisstein's World of Physics*: http://scienceworld.wolfram.com/physics/FundamentalForces.html).

18 *Wolfgang Pauli. Writings on Physics and Philosophy*, edited by C. P. Enz and K. Von Meyenn (Springer, Berlin, 1994).

19 *Wonder Boys* by Michael Chabon (Fourth Estate, London, 2008).

20 *The God Particle: If the Universe Is the Answer, What Is the Question?* by Leon Lederman and Dick Teresi (Mariner Books, Wilmington, 2006).

21 'Discovery or Manufacture?' Tarner Lecture, 1938. Reprinted in *The Philosophy of Physical Science* by Arthur Eddington (University of Michigan Press, Ann Arbor, 1958).

22 'The Reines–Cowan Experiments: Detecting the Poltergeist': http://permalink.lanl.gov/object/tr?what=info:lanl-repo/lareport/LA-UR-97-2534-02.

23 'A Proposed Experiment to Detect the Free Neutrino' by F. Reines and C. L. Cowan, Jr (*Physical Review*, vol. 90, 1 May 1953, p. 492).

24 KSU Physics Ernest Fox Nichols Lecture by Herald Kruse, 1 March 2010: https://www.phys.ksu.edu/alumni/nichols/2010/kruse-lecture.pdf.

25 Ibid.

26 'Detection of the Free Neutrino: A Confirmation' by C. L. Cowan, Jr, F. Reines, F. B. Harrison, H. W. Kruse and A. D. McGuire (*Science*, vol. 124, 20 July 1956, p. 103). 'The Neutrino' by Frederick Reines and Clyde Cowan, Jr (*Nature*, vol. 178, p. 446).

27 Unfortunately, Clyde Cowan died in 1974 and so did not share the Nobel Prize with Frederick Reines.

28 Possibly, Reines had confused events with an earlier episode with a drunken Pauli. In late 1953, when news had reached Pauli of Reines and Cowan's first faint hint of the existence of the neutrino at Hanford, he and his friends finished their dinner and climbed the Uetliberg mountain above Zurich. On the way down later that evening, Pauli, wobbly from the wine he had drunk at the meal, had to be supported on both sides by his friends to avoid falling over (*No Time to Be Brief: A Scientific Biography of Wolfgang Pauli* by Charles Enz (Oxford University Press, 2010)).

29 KSU Physics Ernest Fox Nichols Lecture by Herald Kruse, 1 March 2010: https://www.phys.ksu.edu/alumni/nichols/2010/kruse-lecture.pdf.

30 For the discovery of the solar neutrino problem, Ray Davis shared the 2002 Nobel Prize in Physics.

31 The proof that neutrinos oscillate between three 'flavours' as they fly through space and therefore have mass was obtained independently by Takaaki Kajita at the Super-Kamiokande detector in Japan and Arthur McDonald at the Sudbury Neutrino Observatory in Canada. They shared the 2015 Nobel Prize in Physics.

32 So John Updike was incorrect to say, in the poem quoted at the start of this chapter, that neutrinos 'have no charge and have no mass', and he was also wrong to say that neutrinos 'do not interact at all' since they interact (admittedly rarely) via the weak nuclear force, and of course via gravity. But I like the poem!

33 One of the most amazing images in the history of science was created by the Super-Kamiokande neutrino detector, deep underground in the Japanese Alps. It is an image of the Sun taken at night, not looking up at the sky but down through 12,700 kilometres of the Earth, and generated not by light but by neutrinos. If there is ever an illustration of how, to neutrinos, the Earth is the most rarefied of fogs, it is that image.

34 If there were many more than three generations of neutrinos, the gravity of their extra mass would have braked the expansion of the Big Bang fireball, causing the universe to stay denser and hotter for longer, so that nuclear reactions forged a different amount of helium than astronomers observe. It is possible to have more than three types, however, if they are of a type known as 'sterile'. The normal neutrinos, although antisocial, do interact with normal matter occasionally via nature's 'weak nuclear force'. Sterile neutrinos would not even do this; their sole interaction with normal matter would be via the gravitational force.

35 'Neutrinos Suggest Solution to Mystery of Universe's Existence' by Katia Moskvitch (*Quanta Magazine*, 12 December 2017: https://www.quantamagazine.org/neutrinos-suggest-solution-to-mystery-of-universes-existence-20171212/).

36 *Neutrino* by Frank Close (Oxford University Press, Oxford, 2010).

6: The day without a yesterday

1 'Extra-Terrestrial Relays' by Arthur C. Clarke (*Wireless World*, October 1945, p. 305).

2 The switching and the subtraction are done electronically.

3 Incredibly, it was possible to post a baby in the mail in the US until 1913 ('A Brief History of Children Sent Through the Mail' by Danny Lewis (Smithsonian.com, 14 June 2016: https://www.smithsonianmag.com/smart-news/brief-history-children-sent-through-mail-180959372/)).

4 An unlikely friend of George Gamow's was the English quantum theorist Paul Dirac. Gamow liked to talk and Dirac was happy to listen, and the garrulous Gamow even taught his taciturn friend to ride a motorbike.

5 The term 'Big Bang' was coined in 1949 by Fred Hoyle, who, ironically, never believed in it.

6 Fred Hoyle would later discover that the route to building heavier elements involved three helium nuclei colliding to form a nucleus of carbon. This highly unlikely 'triple-alpha' process was significant inside stars because they maintained high densities and temperatures not simply for ten minutes but for millions and even billions of years.

7 'The Origin of the Chemical Elements' by Ralph Alpher, Hans Bethe and George Gamow (*Physical Review*, vol. 73, 1948, p. 803).

8 Ralph Alpher's son Victor writes that Herman had disappointed Gamow by refusing to change his name to 'Delta' ('The History of Cosmology as I Have Lived Through It' by Victor Alpher (*Radiations*, vol. 15, issue 1, spring 2009, p. 8)).

9 If the universe were shrunk by a factor of eight, the energy density of the matter particles would go up by a factor of eight. However, the photons would double their energy, so the energy density of radiation would go up by a factor of sixteen. So, even though today we live in a matter-dominated universe, in the past radiation would have been important. In fact, during the first few hundred thousand years of its existence, the universe was radiation-dominated.

10 The term 'black body' is unfortunate, since it refers to the spectrum off a bright fireball. However, there is method in physicists' madness. A black body is an idealised body that absorbs all photons that fall on it and radiates nothing, hence its blackness. Inside the body, those photons bounce around, sharing their total energy and achieving a black body spectrum. Of course, to observe the spectrum a small hole would have to be made in the body to let out some of the light.

11 'Evolution of the Universe' by R. A. Alpher and R. C. Herman (*Nature*, vol. 162, 13 November 1948, p. 774).

12 Dicke believed in the existence of relic heat radiation in the universe for the opposite reason to Gamow, Alpher and Herman. Rather than the universe beginning in a one-off Big Bang, he subscribed to the idea of a universe swelling and contracting throughout eternity like a giant beating heart. Such

an 'oscillating universe' sidestepped the awkward 'What happened before the Big Bang?' question but had another problem. In 1957, Fred Hoyle and his co-workers had succeeded where Gamow had failed in finding a furnace in which elements heavier than helium could be forged: stars. But if the universe began as hydrogen, and stars then cooked some of it into heavy elements, what had happened to the heavy elements that had been made during the universe's previous cycle of expansion and collapse? There must be a process that destroyed all the universe's heavy elements between the big crunch at the end of a phase of contraction and the Big Bang at the start of the next expansion, and Dicke realised that extreme heat would do the job. During its compression, the universe must have been very hot – many billions of degrees. At such a temperature, the heavy elements would have been slammed together so violently that they would have disintegrated into hydrogen, erasing all traces of the previous era of cosmic history. An unavoidable consequence of such a primordial fireball phase was fireball radiation. Dicke, like Gamow, concluded that the early universe must be pervaded by leftover heat.

13 It took a little more evidence to prove the Big Bang theory beyond doubt. It was necessary, for instance, to measure the afterglow of creation at different frequencies to show that it did indeed conform to a black body spectrum. And it was necessary to observe the distant (and therefore early) universe. Such observations in the early 1960s revealed 'quasars', which no longer exist in today's universe. They confirmed the Big Bang prediction that we live in a changing universe and not an unchanging one, as predicted by the rival 'steady-state' theory of Fred Hoyle, Tommy Gold and Hermann Bondi.

14 'A Measurement of Excess Antenna Temperature at 4,080 Megacycles per Second' by Arno Penzias and Robert Wilson (*Astrophysical Journal*, vol. 142, July 1965, p. 419).

15 Actually, the cosmic background radiation had been both predicted and discovered before it was discovered. Not only had Alpher and Herman predicted it seventeen years earlier in 1948 but a decade earlier than that, in 1938, Walter Adams, using the biggest telescope in the world – the giant 100-inch reflector on Mount Wilson – had noticed something puzzling. Out in the cold of space, tiny dumbbell-shaped molecules of cyanogen were spinning faster than they should. Canadian astronomer Andrew McKellar suggested that they were being buffeted by something – radio waves at a few degrees above absolute zero. With the discovery of the cosmic background radiation, which permeates every pore of the universe, it suddenly became obvious what that 'something' was.

16 *The First Three Minutes* by Steven Weinberg (Basic Books, New York, 1993).

17 Ibid.

7: The holes in the sky

1 *The Hitchhiker's Guide to the Galaxy* by Douglas Adams (William Heinemann, London, 1995).

2 In 1974, Louise Webster returned to Australia to work at the Anglo-Australian Telescope at Siding Spring Observatory. She married a British radio astronomer called Tony Turtle, but sadly, despite having the first-ever liver transplant in Australia, she died at only forty-nine. See http://asa.astronomy.org.au/profiles/Webster.pdf.

3 'Optical Identification of Cygnus X-1' by Paul Murdin and Louise Webster (*Nature*, vol. 233, 10 September 1971, p. 110).

4 The mass of HDE 226868 today is estimated to be twice the average mass estimated for such a star in 1971. Consequently, the black hole in Cygnus X-1 is known to be about fifteen times the mass of the Sun. Since black holes result from the implosion of the core of a massive star in a 'supernova' that blows 90 per cent of a star's material into space, the precursor star must have been a monster of at least 150 solar masses.

5 'Oral Histories – Martin Schwarzschild' (*American Institute of Physics*, 10 March 1977; https://www.aip.org/history-programs/niels-bohr-library/oral-histories/4870-1).

6 Ibid.

7 Although pemphigus vulgaris is still incurable, its symptoms can be controlled with a combination of medicines that stop the immune system attacking the body. Most people start with high doses of steroids, which help stop new blisters forming and allow existing ones to heal. Gradually, the dose is reduced and another medication that reduces the activity of the immune system is used. If the symptoms do not return, it may be possible to stop taking the medication, but many people need ongoing treatment to prevent flare-ups.

8 From *The Prelude, Book Three* by William Wordsworth:
> The antechapel where the statue stood
> Of Newton with his prism and silent face,
> The marble index of a mind for ever
> Voyaging through strange seas of Thought, alone.

9 *Masters of the Universe: Conversations with Cosmologists of the Past* by Helge Kragh (Oxford University Press, Oxford, 2014).

10 In a way, it was good fortune that Erwin Freundlich did not get to observe the total eclipse of 21 August 1914 because Einstein's prediction of the deflection of starlight by the Sun was wrong – only half the value that would be predicted by his final theory of gravity of November 1915.

11 Einstein's special theory of relativity of 1905 had shown that space and time are aspects of the same seamless entity: space–time. As Einstein's mathematics professor Hermann Minkowski said, addressing the eightieth Assembly of German Natural Scientists and Physicians on 21 September 1908: 'The views of space and time which I wish to lay before you have sprung from the soil of experimental physics, and therein lies their strength. They are radical. Henceforth space by itself, and time by itself, are doomed to fade away into mere shadows, and only a kind of union of the two will preserve an independent reality.'

12 This succinct summary of Einstein's theory of gravity is due to the American physicist John Wheeler.

13 'Oral Histories – Martin Schwarzschild' (*American Institute of Physics*, 10 March 1977).

14 Einstein's field equations of gravity actually contain 4 × 4 tables of numbers, which means there are sixteen equations. However, he was able to use 'symmetry arguments' to reduce the number of equations down to ten.

15 *Karl Schwarzschild: Collected Works*, edited by H. Voigt (Springer, Berlin, 1992).

16 Letter from Karl Schwarzschild to Albert Einstein, dated 22 December 1915 ('Collected Papers of Albert Einstein', vol. 8a, document 169).

17 *Einstein and the History of General Relativity*, edited by Don Howard and John Stachel (Birkhäuser, Boston, 1989, p. 213).

18 'Schwarzschild and Kerr Solutions of Einstein's Field Equation – An Introduction' by Christian Heinicke and Friedrich Hehl (7 March 2015: https://arxiv.org/pdf/1503.02172.pdf).

19 *Masters of the Universe: Conversations with Cosmologists of the Past* by Helge Kragh (Oxford University Press, Oxford, 2014).

20 'Cygnus X-1 – A Spectroscopic Binary with a Heavy Companion?' by Louise Webster and Paul Murdin (*Nature*, vol. 235, 1972, p. 37).

21 The discovery of the black hole in Cygnus X-1 was made independently and pretty much simultaneously by American astronomer Tom Bolton at the University of Toronto's David Dunlap Observatory. His paper was published a few weeks after that of Murdin and Webster. 'Identification of Cygnus X-1 with HDE 226868' by Tom Bolton (*Nature*, vol. 235, 4 February 1972, p. 271).

22 I am writing this account in large part because in 1972, aged twelve, I went with my dad to a Junior Astronomical Society meeting at Caxton Hall in London. The subject was the black hole candidate Cygnus X-1 and the speaker was Paul Murdin. It blew my mind!

23 In 1963, the New Zealand physicist Roy Kerr had found a solution of Einstein's theory of gravity for the space–time warpage of a spinning black hole.

24 See *Quantum Theory Cannot Hurt You: Understanding the Mind-Blowing Building Blocks of the Universe* by Marcus Chown (Faber, London, 2008).

25 'Galactic Explorer Andrea Ghez' by Susan Lewis (*NOVA*, 31 October 2006: http://www.pbs.org/wgbh/nova/space/andrea-ghez.html).

26 John Wheeler is often credited with coining the term 'black hole', but he merely popularised it. 'In the fall of 1967, [I was invited] to a conference . . . on pulsars,' he wrote. 'In my talk, I argued that we should consider the possibility that the center of a pulsar is a gravitationally completely collapsed object. I remarked that one couldn't keep saying "gravitationally completely collapsed object" over and over. One needed a shorter descriptive phrase. "How about black hole?" asked someone in the audience. I had been searching for the right term for months, mulling it over in bed, in the bathtub, in my car, whenever I had quiet moments. Suddenly this name seemed exactly right. When I gave a more formal Sigma Xi-Phi Beta Kappa lecture . . . on December 29, 1967, I used the term, and then included it in the written version of the lecture published in the spring of 1968' (*Geons, Black Holes and Quantum Foam* by John Wheeler (W. W. Norton, New York, 2000, p. 296)).

27 'First M87 Event Horizon Telescope Results: The Shadow of the Supermassive Black Hole' by the EHT Collaboration (*Astrophysical Journal Letters*, vol. 875, no. 1, 10 April 2019).

8: The god of small things

1 'Life Is a Braid in Spacetime' by Max Tegmark (*Nautilus*, 9 January 2014: http://nautil.us/issue/9/time/life-is-a-braid-in-spacetime).

2 'The Tyger' by William Blake (*Songs of Experience*, 1794).

3 *Massive: The Hunt for the God Particle* by Ian Sample (Virgin Books, London, 2010).

4 'Broken Symmetries, Massless Particles and Gauge Fields' by P. W. Higgs (*Physics Letters*, vol. 12, 13 September 1964, p. 132).

5 'Peter Higgs in Conversation with Graham Farmelo' at the Centre for Life, Newcastle, 1 November 2016: https://www.youtube.com/watch?v=LZh15QK_TFg.

6 *A Children's Picture-Book Introduction to Quantum Field Theory* by Brian Skinner (https://www.ribbonfarm.com/2015/08/20/qft/).

7 Thanks to Jon Butterworth for this analogy.

8 *A Zeptospace Odyssey* by Gian Francesco Giudice (Oxford University Press, Oxford, 2010).

9 Higgs' inspiration for the mechanism by which spontaneous symmetry-breaking endows particles with masses had come from the phenomenon

of superconductivity, in which a metal cooled to close to absolute zero ($-273°C$) loses all resistance to the flow of an electrical current. The American physicist Philip Anderson had pointed out that inside a superconductor, the collective field from all the particles breaks the symmetry of electromagnetism, giving the photon a longitudinal oscillation and so an effective mass. Because photons have a mass, a magnetic field – which, according to quantum field theory, is composed of photons – has a short range and can penetrate only a short way into the superconductor. This 'Meissner effect' is a perfect analogue of how the Higgs field breaks the symmetry of a gauge theory and gives the massless force carriers a short range. The insight that something like the superconductor mechanism might operate elsewhere in physics and, in particular, be responsible for giving gauge particles mass came from Yoichiru Nambu – the Japanese–American physicist was a great influence on Higgs' thinking.

10 'Conceptual Foundations of the Unified Theory of Weak and Electromagnetic Interactions' by Steven Weinberg (Nobel Lecture, 8 December 1979: https://www.nobelprize.org/prizes/physics/1979/weinberg/lecture/).

11 'Broken Symmetries and the Masses of the Gauge Bosons' by P. W. Higgs (*Physical Review Letters*, vol. 13, no. 16, 19 October 1964).

12 'Evading the Goldstone Theorem' by Peter Higgs (Nobel Lecture, 8 December 2013: https://www.nobelprize.org/prizes/physics/2013/higgs/lecture/).

13 Einstein showed that whether you see an electric or magnetic field depends entirely on your velocity, revealing that neither is fundamental and in fact both are aspects of a single electromagnetic field.

14 'A Theory of the Fundamental Interactions' by Julian Schwinger (*Annals of Physics* 2, 1956, p. 407).

15 In the Standard Model, the strong force results from a gauge theory based on an unbroken $SU(3)$ symmetry called quantum chromodynamics, whereas the weak and electromagnetic forces arise from a gauge theory based on a broken $SU(2) \times U(1)$ symmetry.

16 If Englert's collaborator Robert Brout had not died on 3 May 2011, he would probably have shared the Nobel Prize in Physics.

17 'Facts and Figures about the LHC': https://home.cern/resources/faqs/facts-and-figures-about-lhc.

18 *The Infinity Puzzle* by Frank Close (Oxford University Press, Oxford, 2013, p. 342).

19 The W+, W– and Z0 particles were discovered at CERN's 'Super Proton–Antiproton Synchrotron' in the early 1980s. Weighing 80.4, 80.4 and 91.2 times the mass of a proton respectively, each was almost as massive as an

atomic nucleus of silver. Carlo Rubbia and Simon van der Meer won the 1984 Nobel Prize in Physics for the discovery.

9: The voice of space

1 Indirect evidence of the existence of gravitational waves was actually discovered in 1975. It came from the 'binary pulsar', PSR B1913+16, a system in which two super-compact 'neutron stars' are spiralling together. Careful observations by Russell Hulse and Joseph Taylor revealed that the stars lose orbital energy at exactly the rate expected if they are radiating gravitational waves. For their discovery, the two American astronomers won the 1993 Nobel Prize in Physics.

2 To be precise, the gravitational force between an electron orbiting a proton in an atom of the lightest element, hydrogen, is about 10^{40} times weaker than the electromagnetic force between the particles.

3 'On Gravitational Waves' by Albert Einstein and Nathan Rosen (*Journal of the Franklin Institute*, vol. 223, issue 1, January 1937, p. 43).

4 I am conflating things here, but the story of Einstein asking to use a bulldozer is true. A family friend, Loretta Donato, tells of her uncle working on a building site in Princeton. 'For several days a little old man was sitting on a bench watching my uncle work,' says Donato. 'One day the old man asked my uncle if he would show him how to use the bulldozer, and my uncle agreed. The little old man was Einstein . . . My uncle taught Einstein how to use a bulldozer. But the family joke is that my uncle taught Einstein at Princeton!'

5 'Einstein Versus the *Physical Review*' by Daniel Kennefick (*Physics Today*, vol. 58, issue 9, 2005, p. 43: https://doi.org/10.1063/1.2117822).

6 'Observation of Gravitational Waves from a Binary Black Hole Merger' by B. P. Abbott et al., LIGO Scientific Collaboration and Virgo Collaboration (*Physical Review Letters*, vol. 116, 11 February 2016, p. 061102).

7 See https://eventhorizontelescope.org.

8 The gravitational signal from the merger of two black holes would have been impossible to predict had it not been for a breakthrough made by South African–Canadian physicist Frans Pretorius in 2005. Although exact 'solutions' to Einstein's equations of gravity are notoriously difficult to obtain, Pretorius defied the odds and found one for two black holes in orbit about each other ('Evolution of Binary Black Hole Spacetimes' by Frans Pretorius (*Physical Review Letters*, vol. 95, 14 September 2005, p. 121101: https://arxiv.org/pdf/gr-qc/0507014.pdf)).

9 Gravity can shrink an interstellar cloud of gas and dust to form a compact star if the cloud can shed its internal heat, since the force of hot gas pushing

outwards stymies gravity. This happens when molecules radiate energy in the form of far-infrared light, which is able to escape a gas cloud. However, molecules consist of heavy atoms such as carbon and oxygen, which have been built up from hydrogen inside stars since the Big Bang 13.82 billion years ago. In the beginning, there were no such molecules; consequently, bigger masses with bigger gravity were required to overcome the internal heat of gas clouds and spawn stars. This is why the first-generation stars would have been giants by today's standards.

10 The gamma rays, after travelling 130 million light years across space from the elliptical galaxy NGC 4993, arrived just 1.7 seconds after the burst of gravitational waves. From this, physicists deduced that the speed of gravitational waves is within one part in a million billion of the speed of light ('GW170817: Observation of Gravitational Waves from a Binary Neutron Star Inspiral' by B. P. Abbott et al. (*Physical Review Letters*, vol. 119, 16 October 2017, p. 161,101)).

11 *Black Hole Blues: And Other Songs from Outer Space* by Janna Levin (The Bodley Head, London, 2016).

12 *The Ascent of Gravity: The Quest to Understand the Force that Explains Everything* by Marcus Chown (Weidenfeld & Nicolson, London, 2017).

10: The poetry of logical ideas

1 *The First Three Minutes: A Modern View of the Origin of the Universe* by Steven Weinberg (Basic Books, New York, 1993).

2 *The Assayer (Il Saggiatore)* by Galileo Galilei (1623).

3 *The Cosmic Code: Quantum Physics as the Language of Nature* by Heinz Pagels (Dover Publications, New York, 2012).

4 'The Unreasonable Effectiveness of Mathematics in the Natural Sciences' by Eugene Wigner, in *The Collected Works of Eugene Wigner, Volume VI*, edited by Jagdish Mehra (Springer Verlag, Berlin, 1995).

5 'Geometry and Experience', an expanded form of an address by Albert Einstein to the Prussian Academy of Sciences in Berlin, 27 January 1921.

6 'Symmetry and Currents in Particle Physics' by Murray Gell-Mann (Nobel Lecture, 11 December 1969: https://www.nobelprize.org/prizes/physics/1969/ceremony-speech/).

7 Mathematics is built from building blocks which mathematicians call 'formal systems'. There are a large number of such systems, such as 'Boolean algebra' and 'group theory'. A formal system consists of a set of givens, or 'axioms', and the consequences, or 'theorems', that can be deduced from them by applying the rules of logic. For instance, the axioms of Euclidean geometry

include the statement that 'parallel lines never meet', while the theorems that can be deduced from the axioms include such statements as 'the internal angles of a triangle always add up to 180 degrees'.

8 *Our Mathematical Universe: My Quest for the Ultimate Nature of Reality* by Max Tegmark (Penguin, London, 2015).

9 *A Zeptospace Odyssey: A Journey into the Physics of the LHC* by Gian Francesco Giudice (Oxford University Press, Oxford, 2010).

10 *The Logic of Modern Physics* by Percy Bridgman (Macmillan, New York, 1927).

11 *Conjuring the Universe: The Origins of the Laws of Nature* by Peter Atkins (Oxford University Press, Oxford, 2018).

Further Reading

1: Map of the invisible world

The Neptune File: A Story of Astronomical Rivalry and the Pioneers of Planet Hunting by Tom Standage (Penguin, London, 2000).

The Hunt for Vulcan: How Albert Einstein Destroyed a Planet and Deciphered the Universe by Thomas Levenson (Random House, London, 2015).

The Hunt for Planet X: New Worlds and the Fate of Pluto by Govert Schilling (Copernicus, Berlin, 2008).

The Ascent of Gravity: The Quest to Understand the Force that Explains Everything by Marcus Chown (Weidenfeld & Nicolson, London, 2017).

2: Voices in the sky

Faraday, Maxwell and the Electromagnetic Field: How Two Men Revolutionized Physics by Nancy Forbes and Basil Mahon (Prometheus Books, New York, 2014).

The Man Who Changed Everything: The Life of James Clerk Maxwell by Basil Mahon (John Wiley, Chichester, 2003).

Electric Universe: How Electricity Switched on the Modern World by David Bodanis (Abacus, London, 2005).

Faraday: The Life by James Hamilton (HarperCollins, London, 2002).

3: Mirror, mirror on the wall

The Strangest Man: The Hidden Life of Paul Dirac, Quantum Genius by Graham Farmelo (Faber & Faber, London, 2010).

Antimatter by Frank Close (Oxford University Press, Oxford, 2007).

It Must Be Beautiful: The Great Equations of Modern Science, edited by Graham Farmelo (Granta Books, London, 2002).

4: Goldilocks universe

Home Is Where the Wind Blows: Chapters from a Cosmologist's Life by Fred Hoyle (University Science Books, California, 1994).

Fred Hoyle: A Life in Science by Simon Mitton (Aurum, London, 2005).

The Magic Furnace: The Search for the Origins of Atoms by Marcus Chown (Vintage, London, 2000).

5: Ghost busters

No Time to Be Brief: A Scientific Biography of Wolfgang Pauli by Charles Enz (Oxford University Press, Oxford, 2010).

Inward Bound: Of Matter and Forces in the Physical World by Abraham Pais (Oxford University Press, Oxford, 1988).

Neutrino by Frank Close (Oxford University Press, Oxford, 2010).

8: The god of small things

Smashing Physics: Inside the World's Biggest Experiment by Jon Butterworth (Headline, London, 2014).

The Infinity Puzzle: The Personalities, Politics, and Extraordinary Science Behind the Higgs Boson by Frank Close (Oxford University Press, Oxford, 2013).

A Zeptospace Odyssey: A Journey into the Physics of the LHC by Gian Francesco Giudice (Oxford University Press, Oxford, 2010).

In Search of the Ultimate Building Blocks by Gerard 't Hooft (Cambridge University Press, Cambridge, 1998).

Massive: The Hunt for the God Particle by Ian Sample (Virgin Books, London, 2010).

It Must Be Beautiful: Great Equations of Modern Science, edited by Graham Farmelo (Granta Books, London, 2002).

The Particle at the End of the Universe: The Hunt for the Higgs and the Discovery of a New World by Sean Carroll (Oneworld, London, 2012).

Beyond the God Particle: If the Universe Is the Answer, What Is the Question? by Leon Lederman and Christopher Hill (Prometheus Books, New York, 2013).

Dreams of a Final Theory: The Scientist's Search for the Ultimate Laws of Nature by Steven Weinberg (Vintage, London, 1994).

9: The voice of space

Ripples in Spacetime: Einstein, Gravitational Waves, and the Future Astronomy by Govert Schilling (Harvard University Press, Cambridge, 2017).

Einstein in Berlin by Thomas Levenson (Bantam Books, New York, 2003).

10: The poetry of logical ideas

Our Mathematical Universe: My Quest for the Ultimate Nature of Reality by Max Tegmark (Penguin, London, 2015).

A New Kind of Science by Stephen Wolfram (Wolfram Media, Illinois, 2002).

The Comprehensible Cosmos: Where Do the Laws of Physics Come From? by Victor Stenger (Prometheus, New York, 2006).

Conjuring the Universe: The Origins of the Laws of Nature by Peter Atkins (Oxford University Press, Oxford, 2018).

Is God a Mathematician? by Mario Livio (Simon & Schuster, New York, 2010).

The Universe Speaks in Numbers: How Modern Maths Reveals Nature's Deepest Secrets by Graham Farmelo (Faber & Faber, London, 2019).

Acknowledgements

My thanks to the following people who helped me directly, inspired me or simply encouraged me during the writing of this book: Karen, Laura Hassan, Felicity Bryan, Rowan Cope, Anne Owen, Nick Humphrey, Michele Topham, Manjit Kumar, Graham Farmelo, Paul Murdin, Michela Massimi, Govert Schilling, Marco Drago, Jon Butterworth, Christine Sutton, Ken Strain, Sheila Rowan and Loretta Donato.

Index